I cannot say I believe that there is life out there. But it is possible. Some of us think it is probable. Our first halting steps into space have shown us other worlds far stranger and more interesting than imagined by the authors of the most exotic fiction.

For the first time in the history of our species, we have devised the tools—unmanned space vehicles and large radio telescopes—to search for extra-terrestrial life. I would be very ashamed of my civilization if, with these tools at hand, we turned away from the cosmos.

"I must go
Where the fleet of stars is anchored and the young star-captains glow."

—James Elroy Flecker

OTHER WORLDS

Bantam Books Toronto/New York/London

Published simultaneously in the United States and Canada.
Printed in the United States of America.

by Carl Sagan

produced by Jerome Agel

At an instant some 15-billion years ago, the cosmic clock began to tick—and the Universe, or at least its present incarnation, was born.

For some 10-billion years the clock ticked away. And then our Sun turned on, the planets formed, life and intelligence evolved.

We know the beginning of this story but not the end.

"Out yonder there is a huge world, which exists independent of us human beings and which stands before us like a great, eternal riddle, at least partially accessible to our inspection and thinking. The contemplation of this world beckons like a liberation."

—Albert Einstein

"Ammonia, Ammonia."

Drawing by R. Grossman; © 1962 The New Yorker Magazine, Inc.

A meadow in the middle

of the sky

If the eons that comprise the lifetime of Earth were compressed into the span of a single year, the origin of life would have occurred at the end of January; the colonization of the land in November; the flourishing of the dinosaurs on December 15; the evolution of mammals on Christmas Day; the development of the first primates near dawn on December 31; and the origin of man at 8 p.m. on New Year's Eve. Recorded history would occupy the last 30 seconds of the last day of such a year.

Human beings are subject to the same evolutionary and environmental pressures that caused the extinction or variation of all previous forms of life. We pass this way but once.

Just as in time we occupy but an instant in an expanse of ages, so in space we inhabit a single mote of dust in an awesome and immense Universe. We are curious, and a little lonely. Is it really possible, in our Milky Way Galaxy of 250-billion stars, in a Universe of perhaps 100-billion galaxies, that our Sun is the only star with an inhabited planet?

In the most ordinary environments in the Universe the building blocks of life are formed. Planets seem to be commonplace and unexceptional companions of stars. The selective advantage of intelligence and technology is great. Is it not likely that innumerable technical civilizations have arisen, evolved—and decayed?

Until recently, such musings were idle. But in the last quarter-second of the cosmic clock we have reached a turning point in the evolution of life. For the first time, beings from Earth are able to search for life elsewhere.

Sophisticated unmanned scientific laboratories, costing a small fraction of the NASA or Defense budgets, can now be landed on other planets. An even grander and more exciting search is now within our grasp. A dozen radio telescopes on our planet are capable of distant interstellar communication. Cornell University's great telescope, in Arecibo, Puerto Rico, will shortly be capable of communicating with an identical radio telescope on another planet anywhere in the Milky Way Galaxy.

Such a search is not easy. Even under the most optimistic assumptions on the likelihood of such communicating civilizations, we would have to examine many hundreds of thousands of stars before we achieved a reasonable probability of detecting intelligent life.

We have so far examined, with negative results, only about 300 stars for possible signals directed our way. The search for such signals—the transmission of which would occupy a minute fraction of the resources of an advanced civilization—has been initiated cautiously and conservatively. But even with radio telescopes of existing design, a major search for extraterrestrial intelligence, occupying many decades, could be mustered.

If we succeed, we may find ourselves plugged into a kind of galactic telephone network: a brilliant intellectual commerce with a magical and dazzling diversity of civilizations on other worlds. Our culture will have been deprovincialized and our four-billion-year-old isolation ended.

If we fail, if we find that in an otherwise lifeless Universe a microbe is a miracle, we will know something of the fragile and unique position not only of man but of intelligent life in the Universe.

Whatever the outcome, such a search will have the most profound implications for our view of ourselves and the Universe in which we live. It is the technological ingenuity of mankind that permits the search; but it is the philosophical and spiritual needs of mankind that motivate it.

"2157836 dash 98632170053798176429763 0, and reverse the charges, please."

Drawing by Alan Dunn;

Some scientists have suggested that there are perhaps a million advanced civilizations in the Milky Way Galaxy. Why, then, is the presence of extraterrestrial intelligence not entirely obvious?

Any society substantially in our technological future—say, millions of years ahead of us—will have prodigious capabilities. They will be as far beyond our powers as we are beyond our Stone Age ancestors.

Perhaps such advanced civilizations are able to move stars about and restructure whole galaxies. In that case their presence, some say, would be apparent by looking up at the sky on a clear night. There are many unexplained astronomical phenomena. Might any of them be the manifestations of extraterrestrial intelligence?

Such civilizations should surely be capable of interstellar space flight. There are no fundamental physical objections to starflight, and some schemes already have been proposed, even for our puny technology, for travel among the stars—at speeds ranging from one-tenth the velocity of light to speeds close to the velocity of light. The Galaxy is roughly 100,000 light-years across. At one-tenth the velocity of light it would take a million years to cross the Galaxy.

If an advanced society began at one locale in the Galaxy and gradually spread, establishing colonies that themselves launched exploratory missions—as occurred on Earth with the Phoenician and Greek city-states, for example—the whole of the Galaxy might be explored and colonized in times considerably shorter than its 10-billion-year history.

Why then do we not see a wholesale reworking of the Galaxy? Why are we not visited daily and obviously by extraterrestrial beings? No one knows the answers to these questions. I only list here a range of proposed explanations:

(1) There are no extraterrestrial civilizations. Either we are the first technical civilization or the chances of a society surviving the emergence of high technology—the critical moment in Earth-history through which we are passing today—are nil.

(2) There are many other societies, but it is technologically infeasible to rework a galaxy or to travel conveniently among the stars by spaceship.

(3) The Galaxy has been reworked and the Earth visited many times, but we are too backward to know it. Perhaps some puzzling and still unexplained astrophysical phenomena are due to extraterrestrial intelligence. Perhaps the Earth is visited by a technology so subtle that we do not recognize it.

(4) Conscious efforts are made by advanced extraterrestrial societies to avoid detection—perhaps because of a galactic ethic forbidding interference in emerging societies. In this view the Earth is something like a wilderness area or a zoo.

Some of these alternatives pose methodological difficulties for the scientist. For example, the idea that there are conscious efforts by a superior civilization to prevent us from determining the answers is not subject to proof. But it is subject to disproof.

The problem might be approached by a careful search of the legends and artifacts of man—although no such evidence has yet emerged. But at our command is a powerful test of great promise—the search via radio telescopes for intelligent signals from the transmitters of distant stellar civilizations. If there are signals, explanations (2) and (3) will be favored; if not, explanations (1) and (4).

"Who's there?"
—*Hamlet,* I. i.

The existence of an advanced civilization is, for us, its most important property. Likewise, the existence of a message from elsewhere is its most important property, independent of the nature of the message.

Any society that can transmit a message across great interstellar distances must necessarily have survived, with few exceptions, for a long time. It must have weathered the growing pains of technological adolescence—overpopulation, chemical and biological pollution, the proliferation of nuclear weapons, and the exhaustion of mineral and fossil fuel resources.

The mere receipt of a message from elsewhere implies that someone has survived technology, that it is possible to grow through this adolescence. The message might even contain detailed information on the avoidance of these catastrophes.

The time-scale of interstellar radio communication —decades to muster a successful search, centuries to engage in a single dialogue—forces us to accept the long view. But one of the major present failings of the governments of the planet Earth is their rejection of the long view and their pursuit of short-term gains while the implications for long-term disasters are ignored.

Those holding office in democracies and totalitarian regimes alike are often concerned only with their maintenance of power. But even Stalin held power for only a quarter of a century, and Soviet economic policy was planned in only five-year increments.

What we need are leaders planning coordinated strategies for the planet Earth on time-scales of decades and centuries and people willing to sacrifice short-term gains to avoid long-term disasters.

COSMIC REPORT CARD

Earth

(By Monitor 4SJ AKA MAN)

F

A—Excellent	**—Recommended for Contact**
B—Good	**—Recommended for Assistance**
C—Satisfactory	**—Continue Observation**
D—Fair	**—Recommended for Modification**
F—Failure	**—Recommended for Termination**

The Rosetta Stone is a message written in three languages—Greek, Egyptian hieroglyphic and demotic—not to provide assistance for future translators, but to insure comprehensibility in a polyglot population.

An interstellar message would be rather different, a conscious attempt to communicate with beings whose native language is not the language of the transmitting civilization. But the transmitting civilization and the receiving civilization will share some things in common; most important, if the message is by radio, they share radio telescopes, radio technology and radio science. Like ham operators on Earth, their first conversations may be about frequencies, band-passes and receivers. We can expect a vast arena of common physics, mathematics, astronomy and chemistry. Science is the Greek of the interstellar Rosetta Stone.

Using this common language, we may be able to comprehend the exotic interstellar hieroglyphics. It is possible to transmit two-, three-, or four-dimensional television pictures via interstellar radio and to indicate clearly that they are television pictures. Once pictures are transmitted, it is extremely simple to develop language—by show-and-tell.

After receipt of the first interstellar message, it will be for us a little like entering kindergarten. We know school is important, we know it is to our advantage, but we are a little afraid. Will we be able to understand what the teacher is saying? Can we continue, for many years, to wrestle with such difficult stuff? Will we be equal to the new social responsibilities of school? We have been taken from our parents and are being taught things they never knew. We will never be the same again.

Just as the first day of school marks the beginning of the end of childhood, the first communication with extraterrestrial intelligence will mark the beginning of the maturation of the human species.

In talking about interstellar contact, I find many people respond, "It sounds perfectly fine to listen, but for heaven's sake don't send. Don't tell them where we are."

But it is too late. We have revealed our presence to the Galaxy by mundane military and commercial activities: radio, radar, television, signals expanding outward in all directions from Earth at the velocity of light. That information is now about 25 light-years out. There are unlikely to be any civilizations as close as 25 light-years. The nearest are probably a few hundred light-years away.

So we may have some breathing room—a few hundred years before we are discovered.

"Frankly, we were kind of hoping it was Donald Duck that was real and the rest of you that were fiction."

"A word once let out of the cage cannot be whistled back again."

—Horace

I see in the Universe enormous and mysterious and exquisite order. But I also see vast chaos: Stars and whole galaxies exploding; turbulent, writhing interstellar clouds; devastated lunar and planetary landscapes. A quasar explosion in the center of a galaxy could destroy life on perhaps a million inhabited planets. I am not sure that in such a picture I see a traditional, omnipotent, omniscient and benevolent being overseeing the welfare of all his creatures.

Why are the laws of nature the way they are? We could well imagine a Universe in which the physical laws were slightly different, and stars, planets and life impossible. It is at least interesting that we live in a Universe that permits stars and planets, life, intelligence and speculations such as these.

Many scientists would say that the question, Why are the laws of nature the way they are?, is a silly question; that there are no experiments that can be performed on this question; that the question is operationally undefined; that it is a waste of time to consider it.

But still, sometimes, I wonder. . . .

"Night's candles are burnt out, . . ."

—Romeo and Juliet, III. v.

There are no reports in Europe of the Crab Nebula supernova outburst of July 4, 1054. This titanic stellar explosion was reported in China, Korea, Japan and America. Why not in Europe?

The Aristotelian dictum that the sky is perfect and unchanging had been incorporated by the Church. Those Europeans were in mortal danger who, between July and September, 1054, recognized a brilliant star that had not been there earlier, a luminary so dazzling it could be seen in full daylight and read by at night. For the Inquisition was just beginning. The safest course was to ignore this evidence of the mutability of the heavens. In Asia and North America there were no religious proscriptions against observing and reporting nature, and there the supernova was noted.

In the same way, it is important for us today to see the world as it really is, and not as we wish it to be. In the Eleventh Century the ecclesiastical authorities believed that they were doing good by advocating under extreme penalty the immutability of the heavens. Today we profess to know better.

But today there are still doctrines advocated by political, social, economic, religious and racial groups as if they were unquestionable truths. We should remember the lesson of science: Everything must be questioned, the point first clearly enunciated by Descartes. It is the essence of the error-correcting aspect of science—and the secret of its success.

"Whoever in discussion
adduces authority
uses not intellect
but memory."

—Leonardo da Vinci

The speed of light (186,000 miles a second) is the ultimate cosmic speed limit—as far as we know. It is not a barrier like the sound barrier, which was merely an engineering challenge for aircraft designers. (The velocity of sound is just the speed at which molecules like to move.)

But if a material object could travel at or beyond the speed of light, then much of what we know about physics is wrong, and a wide range of precision measurements suddenly becomes deeply mysterious. On the other hand, if the velocity of light is an absolute speed limit, then, as Einstein showed, such measurements and much of nature can be understood and predicted in detail.

But the speed of light is not so serious a barrier as some may think. Already proposed are multi-generation spacecraft and metabolic inhibitors to slow down the aging of the crew. Also, where Einstein taketh away he also giveth. If we were to travel very close to the speed of light, time as measured on the spacecraft would slow down as much as we like compared to time as measured on the launch planet. We could fly to anywhere.

There is a fourth proposal, much more speculative, and also arising from Einstein's insights. There are certain places called "black holes" that may be apertures through which we could plunge to other parts of space. But could we plunge through them in less time than it takes to cover the intervening distance in the ordinary way? No one knows.

Thirteen billion years ago, when the Universe
let rip and, in disciplined panic, Creation
spewed
mazy
star-treacle and resin, shrinking balls
of debut fire smoldered and glitched.
Revolving
tantrums
tore themselves in, kernel-tight, then
cooler, began shooting off their planets
like
a brace
of dandelions gone to seed. Even stars don't
live forever; old age makes the outer region
redden
and
swell (Betelgeuse's radiant flush in Orion
is as much a death-spurt as the whale's
flower
of blood).
Like the body, a star grows larger as it runs
down. Neptune once flecked a coral sun
(yellow
now
as a scrub-diamond) that will puff beet-red
in five billion years, and suck the planets
back
to
where they began.

—Diane Ackerman

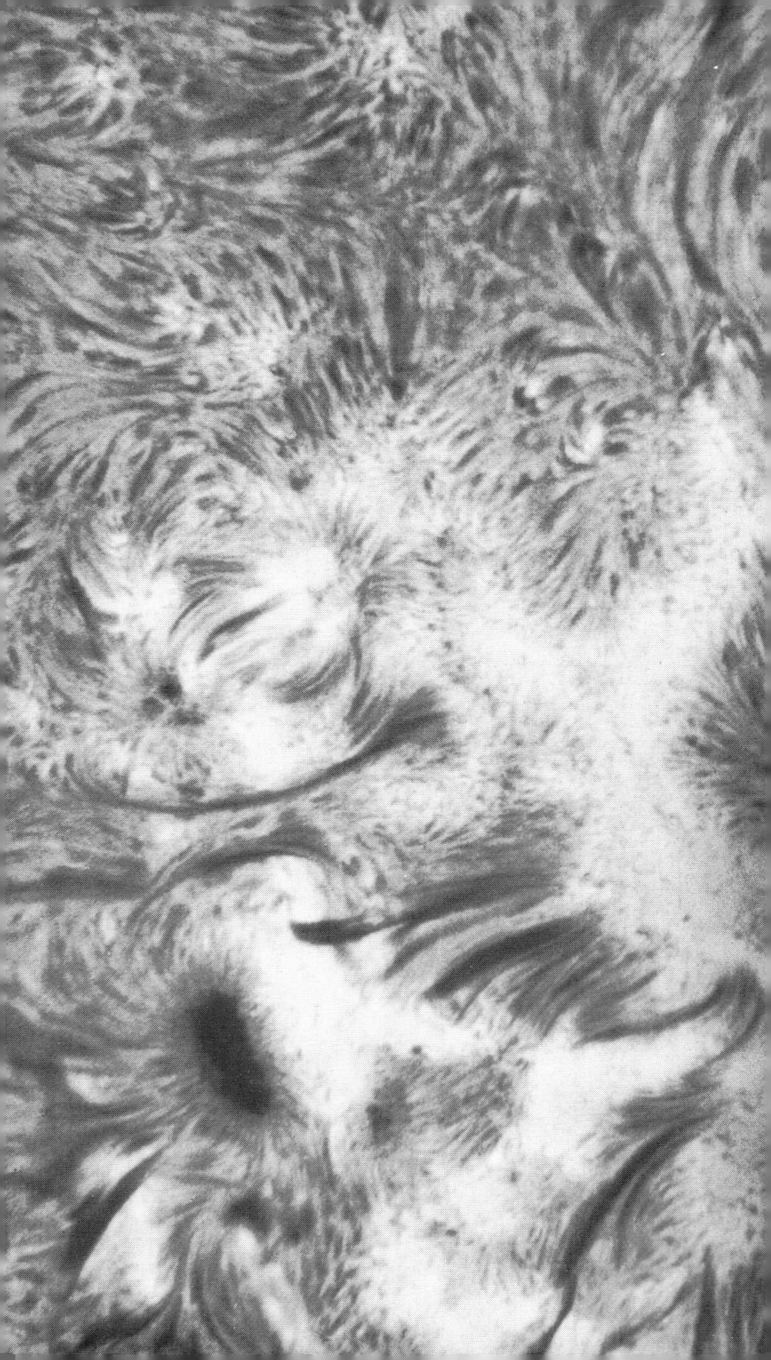

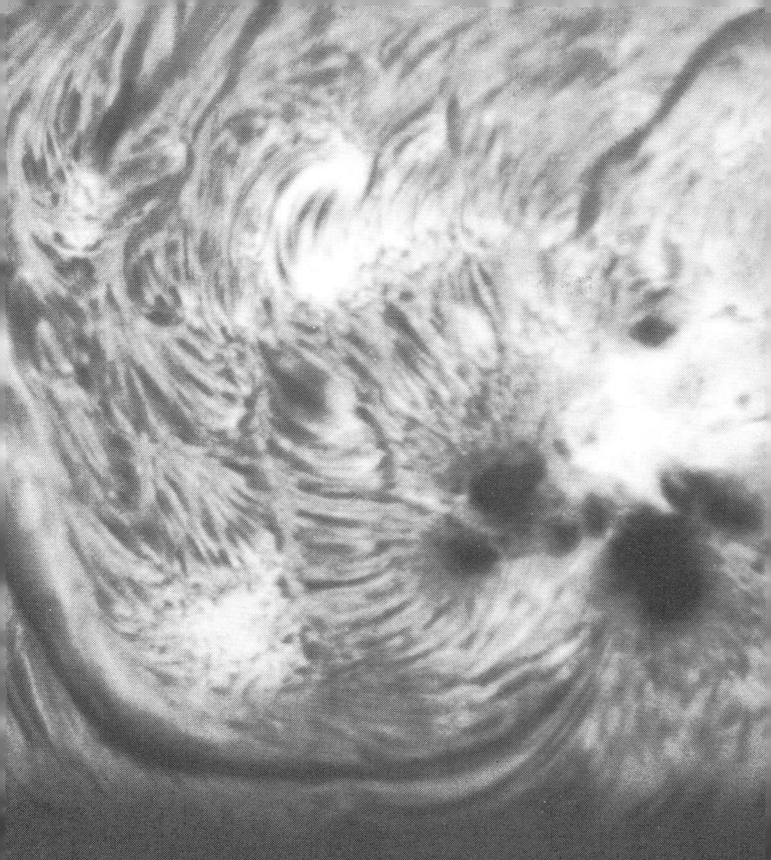

Hob-nobbing
around the Mercurian noon, with only a thin
whiskey-colored veil
to waylay it,
the Sun must burn carte-blanche in the sky,
growing white-hot
and even larger
as the day grows, spitting, burgeoning, and
flaring out
at the seams
like an infra-red sunfish gravid with roe.

—Diane Ackerman

"What is it all but a trouble of ants in

the gleam of a million million of suns?"

—Alfred, Lord Tennyson

"Look at the stars! look, look up at the skies!
O look at all the fire-folk sitting in the air!
The bright boroughs, the circle-citadels there!"

—Gerard Manley Hopkins

"A sad spectacle. If they be inhabited, what a scope for misery and folly. If they be not inhabited, what a waste of space."

—Thomas Carlyle

"In a little while men who will know how to bear the strange gravitations, the altered pressures, the attenuated unfamiliar gases and all the fearful strangeness of space will be venturing out from this earth. This ball will be no longer enough for us; our spirit will reach out . . . Cannot you see how that little argosy will go glittering up into the sky, twinkling and glittering smaller and smaller until the blue swallows it up? They may succeed out there; they may perish, but other men will follow them. . . ."

—H. G. Wells

"Professor Zlata! You're just in time to be the planet Neptune!"

K. E. Tsiolkovskii

"When he shall die,
Take him and cut him out in little stars,
And he will make the face of heaven so fine,
That all the world will be in love with night,
And pay no worship to the garish sun."

—*Romeo and Juliet,* III, ii.

"Wandering between two worlds, one

dead, the other powerless to be born." 

—Matthew Arnold

"So Apollo bore me from the fray."

—Horace

What a pity that our nearest neighbor is so dull. But a person who has never seen another will find even a dull neighbor interesting.

"She, she is dead; she's dead; when thou know'st
this,
Thou know'st how dry a cinder this world is."

—John Donne

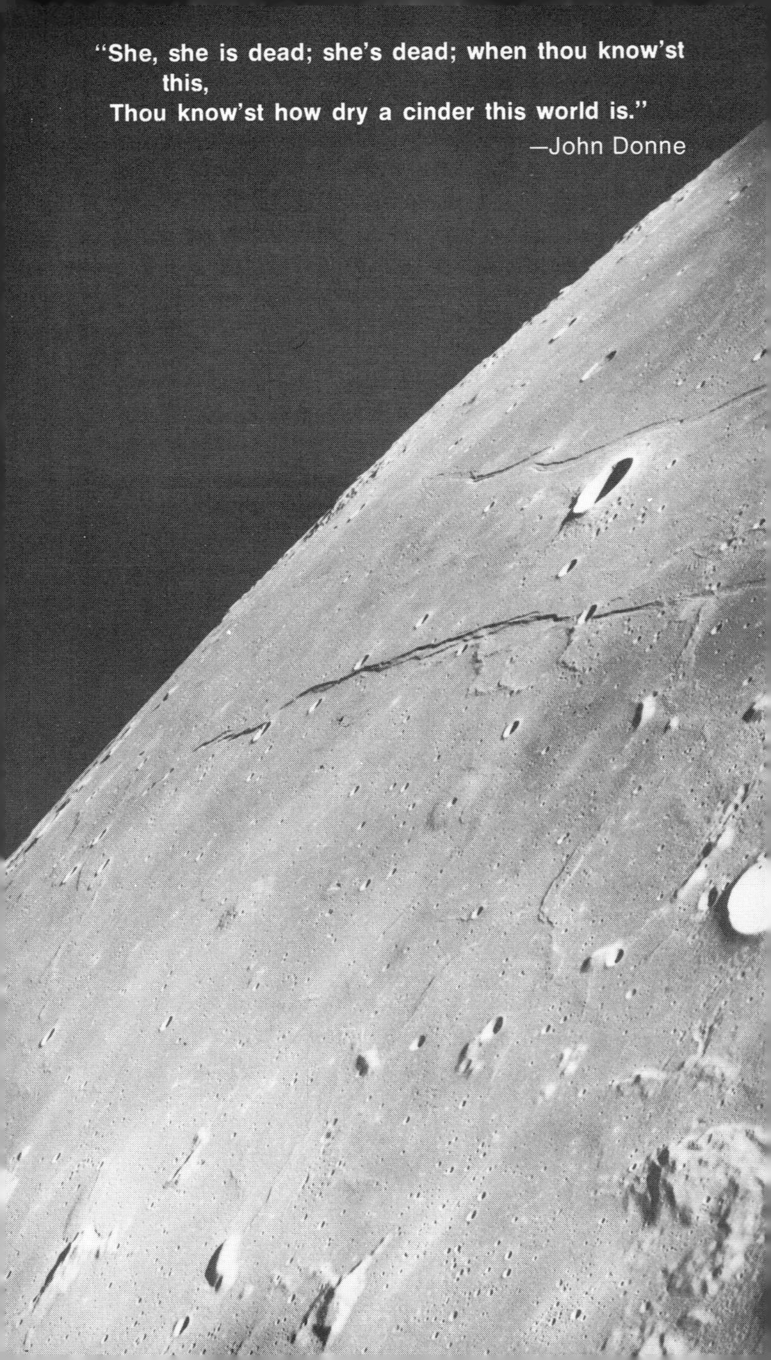

Four-billion years ago the Moon was far from dull. Photographs show clear evidence of a complex series of collisional catastrophes that essentially ended some four-billion years ago, in the final stages of the Moon's formation (and not a few thousand years ago as suggested by Immanuel Velikovsky and his disciples). It is unlikely that the Earth of that remote time would have escaped similar catastrophes. But wind and water and mountain-building have erased these scars from the face of the Earth while they have been meticulously preserved on the airless, waterless, weakly tectonic Moon.

The side of the Moon that faces Earth has a number of large impact basins flooded with lava. The far side has no such features. It is not understood why this asymmetry exists.

Returned lunar samples have shown a remarkably anhydrous surface. Where did all the water go? Seismometers have provided some evidence on the interior structure and early evolution of the Moon. An astonishingly thin and time-variable atmosphere has been uncovered.

When Apollo, an essentially political program, was canceled, the United States was left with no program for the continuing exploration of the Moon. The future study of our natural satellite apparently will be performed only by unmanned orbiters and roving vehicles like the Luna series of the Soviet Union.

LETTERS FROM THE PLANETS.—IV.
"WE PERCEIVED THE CARS OF THE MERCURIANS FLOATING IN SPACE."

FLEET MERCURY (viewed by Mariner 10), the nearest planet to the Sun, is remarkably Moon-like, a world without air or water, sporting the scars of billions of years of impact of space debris. A feature present on Mercury but not on the Moon is a set of polygonal ridges on the periphery of the great Caloris basin, conceivably due to compression in the early stages of the evolution of Mercury and to the presence of a dense iron interior. Despite its slow rotation period of 59 Earth days, Mercury has a detectable magnetic field.

VENUS: The outermost clouds of Venus, photographed in ultraviolet light by Mariner 10. These clouds are higher than the visible clouds of Venus, which are responsible for the great brilliance of the planet as seen from Earth with the naked eye. Some 50 miles below these clouds is the broiling surface of Venus: temperatures of about 900 degrees F.; pressures of about 90 times that at the surface of Earth; and an atmosphere composed largely of carbon dioxide but with traces of hydrofluoric, hydrochloric and sulfuric acid. Why Venus with so dense an atmosphere should have so much less water than Earth is an unsolved problem in planetary physics.

LENOX RAISED THE VISOR OF HIS HELMET.

"WE POISED OUR CAR OVER THE GLITTERING
DOMES AND TOWERS OF THE GREAT CITY
OF THE NORTH POLE."

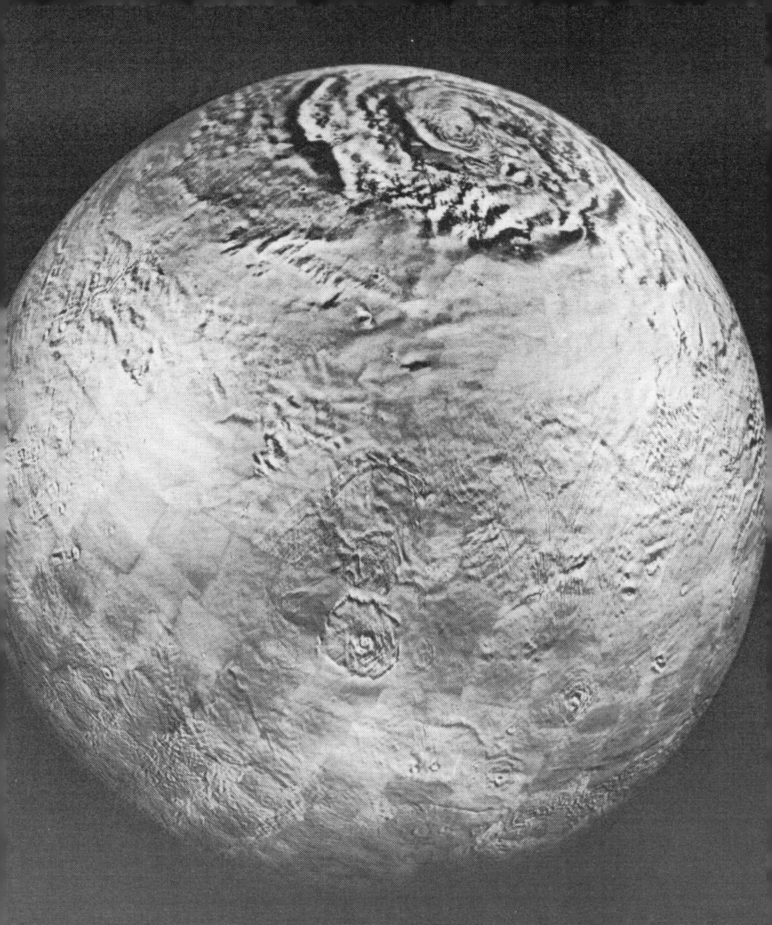

MARS as revealed by Mariner 9 is a spectacularly interesting little planet. It has immense young volcanoes; sinuous channels probably produced in the Martian past by running water; and great dust storms generated by 200-mile-per-hour winds. Mars is a world of craters and towering volcanic summits, of deserts and frozen polar wastelands—a world of great geological and meteorological interest, but a biological question mark. Mars may have a contemporary biology or a dormant biology awaiting the return of more clement conditions. Or, it may be lifeless.

THE MOONS OF MARS: Phobos (right), the inner moon, and Deimos, the outer, are the two natural satellites of Mars. Deimos is about seven miles across in a 12,500-mile-high orbit; Phobos is about 15 miles across in a 4000-mile-high orbit and revolves so fast that it rises in the West and sets in the East. Their cratered and battered faces were first seen during the Mariner 9 mission in 1971-1972. The two largest craters on Phobos are named Hall and Stickney, after the American astronomer and his wife who together were responsible for the telescopic discovery of the two moons in 1877. The two prominent craters on Deimos are named Voltaire and Swift, whose fictional romances included casual guesses at the existence of the moons. Both satellites keep the same face to Mars, as our Moon does to Earth.

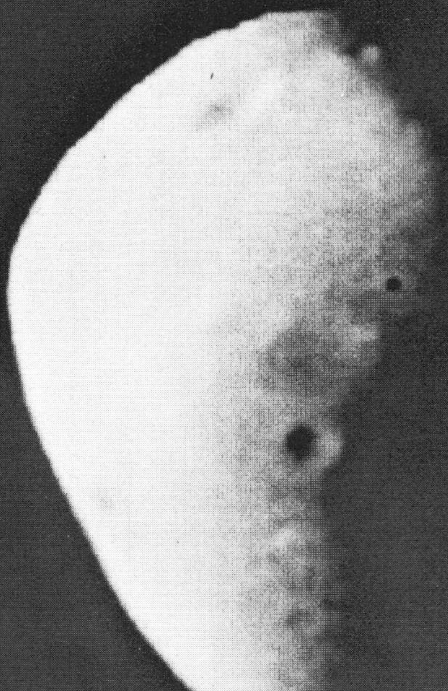

MARS BY NIGHT.

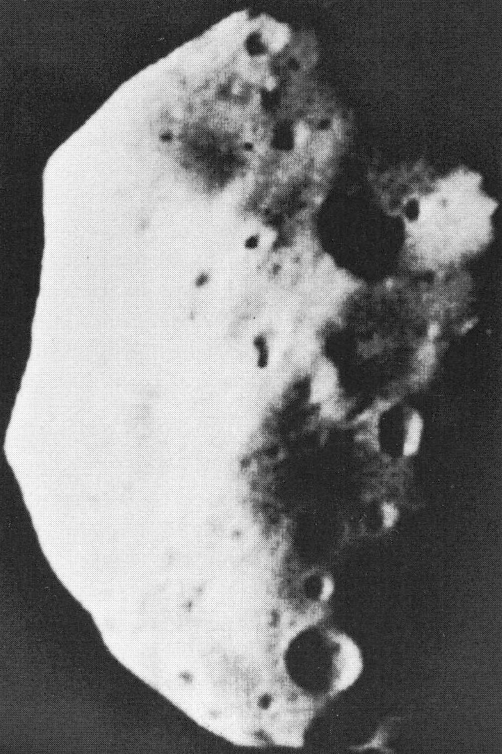

"WHAT A STRANGE SCENE WAS THEN OPENED TO OUR WONDERING EYES!"

JUPITER as seen from Pioneer 10. The bright clouds are made of high-lying ammonia crystals. The darker clouds are at deeper and warmer locales. The surface of Jupiter—if one exists—is at much greater depths than we have seen to. Jupiter contains the most mass and the most angular momentum of all the planets. Its atmospheric composition is similar to that of the primitive Earth at the time of the origin of life, and this giant planet may be the key to many questions on the origin and evolution of the planets as well as, perhaps, to questions on the early history of life on Earth. Jupiter was examined by Pioneer 11 in December, 1974, and will be again by Mariners 11 and 12 in 1979. (After observing Jupiter, Pioneer 10 was accelerated out of the solar system by the gravity of the giant planet. The spacecraft is streaking across interstellar space with a tiny golden message reporting a little of the locale, epoch and nature of its builders.)

"Forward, forward, let us range,
Let the great world spin for ever..."

—Alfred, Lord Tennyson

SATURN: This beautiful ringed planet is in many of its properties similar to Jupiter. It has cloud bands and belts like Jupiter, and like Jupiter it radiates to space approximately twice as much energy as it receives from the Sun. This energy may be due to the slow gravitational contraction of the planet, a vestigial property that it shares with its cousins, the younger stars. The rings of Saturn

are due either to a moon that strayed too close and was torn apart by the tidal gravitational forces of Saturn, or to a ring of material that was prevented by Saturn's gravity from coalescing into a satellite. Recent evidence suggests that the rings are a very thin sheet composed of millions of icy boulders 10 centimeters to a meter across. It is not known why no other planet in the solar system has such a debris ring.

1906 film
The '?' Motorist

TITAN is the largest moon of Saturn—about 3000 miles across—and is the only satellite in the solar system that holds a dense atmosphere. Titan is almost entirely ice—ammonia ice, water ice, but especially methane ice. Its volcanoes may spew out molten lava--not rock lava, but liquid ammonia, water, and methane. Its thick red clouds are probably composed of organic molecules. Pre-biological organic chemistry seems to be in progress. Because of Titan's low gravity, hydrogen gas must be rushing away to space at great speed. But the hydrogen that escapes Titan's gravity cannot escape Saturn's gravity; there should be a doughnut of hydrogen gas surrounding Saturn in Titan's orbit.

Because of its moderately dense atmosphere and low gravity, Titan is the easiest object to land on in the entire solar system. A prime target for future space-vehicle missions, it will be examined close-up in 1981 by the Mariner 11 and 12 flybys. Entry probes may be sent to its enigmatic surface in the late 1980s.

Aqua-blue colossus lurking in the outback,
Neptune wheeled round its solar bevel once
every
160-odd
years: aloof, unheard of, and ill-defined:
a verdigris stopgap in the cosmic pause
hinted
at
but untenable for nearly 2 millenia, then
wrenched from Uranus's mathematical rib.

Neptune
is
elusive as a dappled horse in fog. Pulpy?
Belted? Vapory? Frost-bitten? What we know
wouldn't
fill
a lemur's fist.

—Diane Ackerman

PÉRÉGRINATIONS D'UNE COMÈTE

. . . . that frogspawn cloud of comets
ambling up and down the carborundum sky,
whose sudsy white nougats of dust and ice
may be the only spirits in a lidless night.

—Diane Ackerman

Space is nice. And it's pretty empty. It's as empty as a building 20 miles long

20 miles wide and 20 miles high that contains only a single grain of sand.

"Nymph, in thy orisons
Be all my sins remember'd."

—*Hamlet,* II. ii.

"As I walk'd through the wilderness of this world."
—John Bunyan

A prowling holocaust keeling low in the sky
heads westward
for another milkrun.
When the Sun sets on the Mercurian empire;
it
idles
on each horizon and lurches back, broiling
the same arc
across the sky.
Day in and out. A target gone berserk in a
shooting
gallery.
—Diane Ackerman

The hell of Hieronymus Bosch is not identical to modern views of Venus—but it comes close.

"Father, O father! what do we here
In this land of unbelief and fear?
The Land of Dreams is better far,
Above the light of the morning star."

—William Blake

"I am a little world made cunningly
Of elements, and an angelic sprite."
—John Donne

SUN
EARTH

COME TO BARSOOM

"What shall we tell you? Tales, marvellous tales of ships and stars . . ."

—James Elroy Flecker

Some ideas are considered "far out" for reasons that involve a lack of imagination or nerve—rather than a lack of scientific plausibility. One such idea is the speculation that Mars is inhabited by largish beasts.

Some people are nervous about the possibility of life on Mars, even simple life. There have been so many premature and unsupported speculations on this topic that some scientists and newspaper editorialists often take the most conservative viewpoint they can imagine. They are sometimes found supporting the curious principle: "When faced with two alternatives, in a subject about which we know nothing, choose the least interesting."

Our present understanding is that the contemporary Martian environment is not so severe as to exclude life nor is it so paradisiacal as to guarantee it. Our knowledge of exobiology is simply not advanced enough for us to make such decisions by pure thought. Rather, we must land on the planet and examine it directly.

When asked to picture what will be uncovered by the first few Mars landers, many scientists will admit the possibility of microscopic organisms (microbes), but will deny vehemently the possibility of macroscopic organisms (macrobes)—that is, beasts discernible to an unaided human eye on the surface of the planet. In fact, there is no reason to exclude from Mars organisms ranging in size from ants to polar bears. And there are even reasons why large organisms might do somewhat better than small organisms on Mars.

Mars is, compared to Earth, cold and dry. The heat and water that must be retained by a small organism are proportional to the cube of the characteristic size of the organism. But the rate at which heat and water are lost to the organism depends only on its area, not its volume, and therefore is proportional to the square of its size. Organisms with an interest in the conservation of heat and water thus may select larger sizes. (This is one reason why there are large mammals preferentially in the terrestrial Arctic.)

It is true that on Earth there are often microbes without macrobes, but never vice versa. But Mars, with a climate varying in time between balmy and deep Ice Age, may here and there have macrobes and no free-living microbes.

We must maintain a tolerance for ambiguity for at least a few more years. Then we will be landing on Mars and searching for both little beasts and big ones, Martian microbes and Martian macrobes.

—Editorial, The New York Times, July 30, 1965

The Dead Planet

Mars is probably a dead planet. The astronomers of past decades who thought they detected canals on the Martian surface and speculated that it might have bustling cities and beings engaged in lively commerce were victims of their own fantasies. So, too, were the science-fiction writers who made Mars a realm of epic battles and of weird but fascinating civilizations. The red planet is not only a planet without life now but probably always has been.

Such, tentatively at least, are the conclusions suggested by the remarkable pictures released yesterday—pictures that fully support the impression of a most inhospitable planet given by the earlier reported findings of Mariner 4. Mars, it now appears, is a desolate world of a strangeness almost beyond imagination.

Its surface bathed in deadly radiation from outer space, it has very little atmosphere and has probably never had large bodies of water such as those in which life developed on this planet. And, despite the craters evident in the incredibly clear Mariner 4 pictures, it seems to have n mountains or other topographical features that woul testify to the kind of dynamic instability below the su face that has produced the varied landscape of this pla

A whole host of new sciences is being born—extrate trial geology most obviously among them. By lea more about Mars—even a lifeless Mars—men will stand better the origin of the solar system. And, b ... greater detail ... of wh- ... are the red

Better Red Than Dead

"The solemn peaks but to the stars are known,
But to the stars, and the cold lunar beams:
Alone the sun arises, and alone
Spring the great streams."

—Matthew Arnold

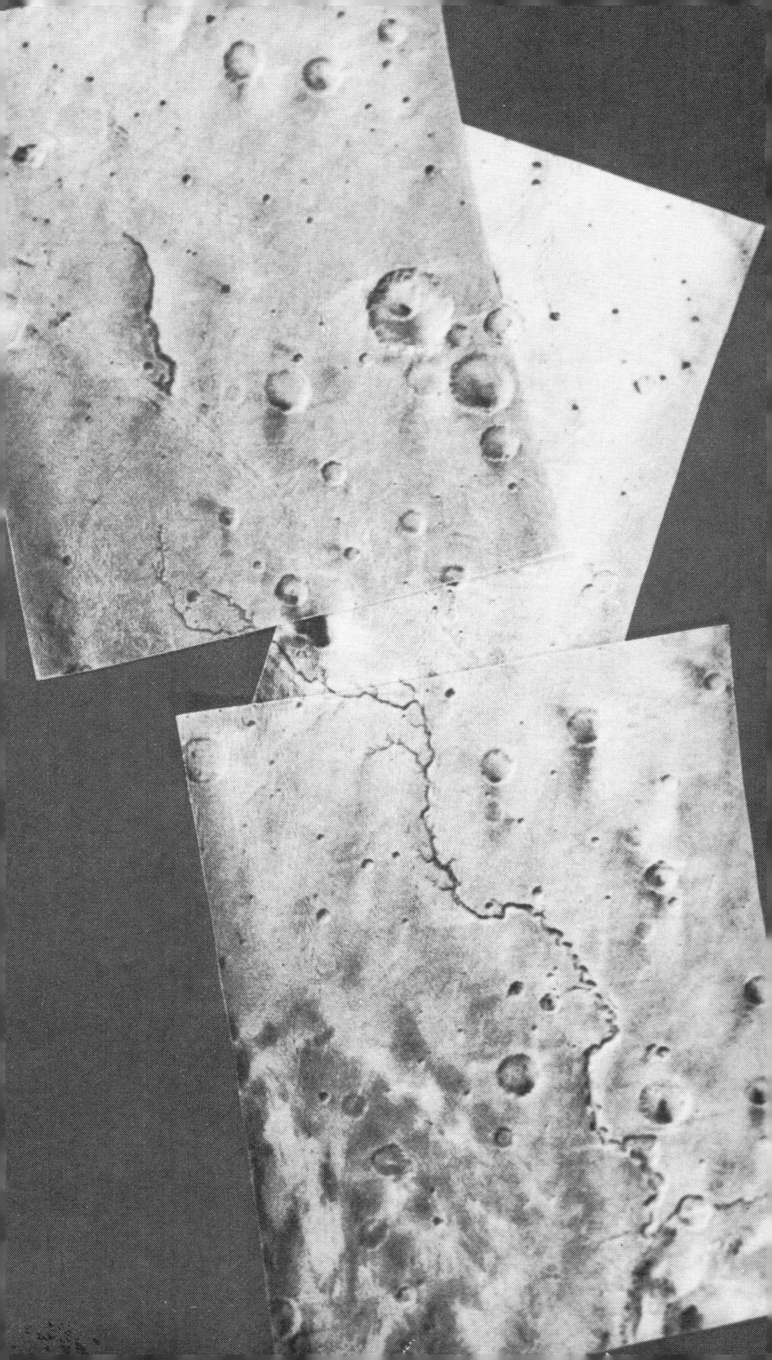

During the Mariner 9 mission, we were stunned to see the planet covered with hundreds of sinuous, tributaried channels running downhill and looking for all the world like dry river valleys on Earth. Since Mars today cannot have liquid water (the total atmospheric pressure is too low), I proposed that we were examining Mars in the midst of a great global Ice Age and that at least once in the past, and probably many times, Mars had higher pressures, somewhat higher temperatures, gurgling brooks and mighty coursing rivers. The great concentration of these channels towards the Martian equator—the only place on the planet where daytime temperatures are above the freezing point of water—seems to argue powerfully that the channels were made by liquid water.

Since there is no way for liquid water to have escaped from Mars (the escape rate from the upper atmosphere is far too low), it today must be sequestered somewhere on the planet. The most likely place is the Martian polar caps, where I believe as much as one Earth atmosphere of carbon dioxide and other gases may be buried.

At Cornell University, my colleagues O. B. Toon and Peter Gierasch and I worked out a detailed theory of climatic change on Mars in which the denser atmosphere comes pouring off the polar cap when the axis of Mars nods more towards the Sun; or when massive quantities of dark equatorial dust lower for a period of a century or more the brightness of the polar caps; or when the Sun substantially increases its brightness.

Imagine my surprise when in the midst of working on explanations of the Martian channels I opened my local morning newspaper, the Ithaca Daily Journal, to find a segment of one of the major channels reproduced with the following caption: "Jeanne Munns, 57, prepares for a spin in her open-cockpit biplane she has had for about two years." There have been many explanations of the channels, but this is surely the most daring. The Munns hypothesis puts our idea of massive climatic variation on Mars in the realm of safe prediction. But I keep wondering if somewhere else the Ithaca Daily Journal has published a photograph of a brave and forthright aviatrix with the caption: "Segment of Martian sinuous channel, attributed by Cornell scientists to running water."

She Loves to Fly

Jeanne Munns, 57, prepares for a spin in her open-cockpit biplane she has had for about two years.

"A brighter Hellas rears its

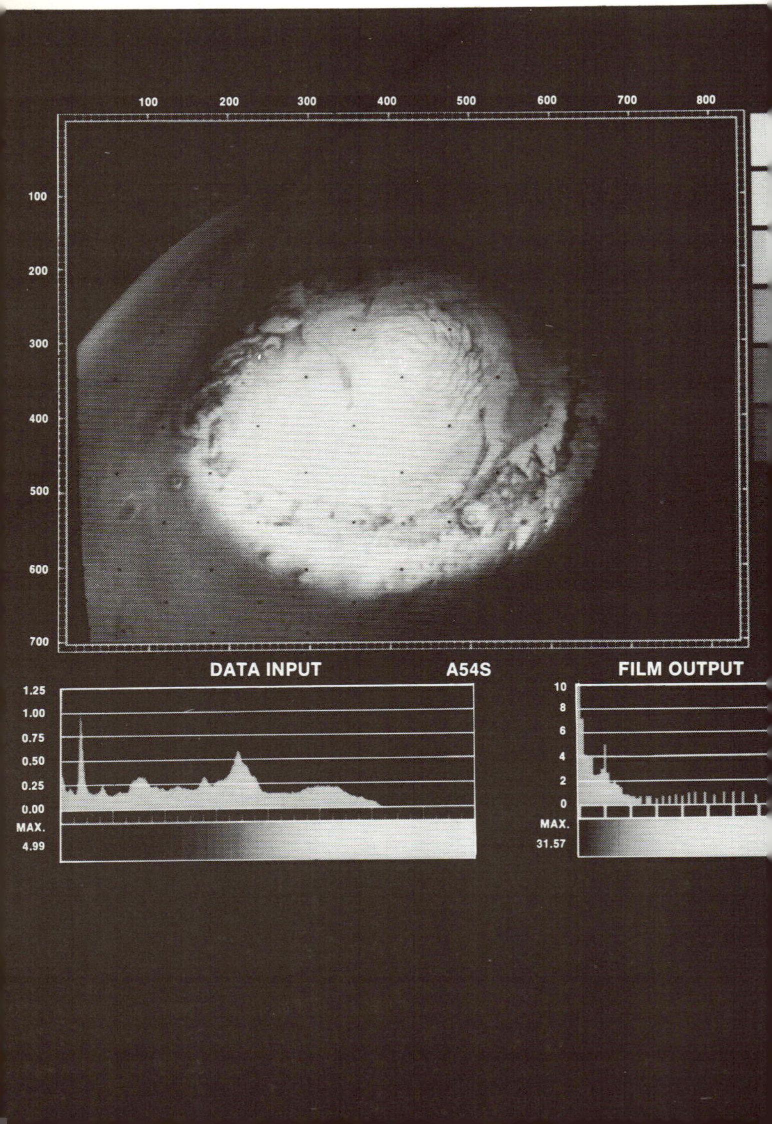

ıountains From waves serener far"

—Percy Bysshe Shelley

MTC-MTVS

MARINER 9 PLYBK 528/529/533
ORBIT , SET
TIME FROM PERIAPSIS 2.20.44
SLANT RANGE 11664 KM
VIEWING ANGLE 39.028 DEG
PHASE ANGLE 56.578 DEG
LIGHTING ANGLE 65.487 DEG
LATITUDE-W LONGITUDE
R1 (UL) 46.525, 65.769
R3 (UR) 52.396, 146.111
R5 (CENTER) 87.871, 330.357
R7 (LL) 56.309, 354.623
R9 (LR) 59.671, 245.519
56.309, 354.623 5
PICTURE 54 220 23.20.47.011
CAMERA A DAS 13028127
FILTER-15 DSN 14
POLAR 60 RATE 4.05KBPS
EXPOSURE-04 PN ERRORS 7
48MSEC PIX SPIKES 435
F700/P000/M000

SHADING CORRECTED

STRETCH CONTROL—AUTO—ES
LOW 063=00 HI 536=77
TRANSLATION—773
MTC 4296-102 07 AUG 1972

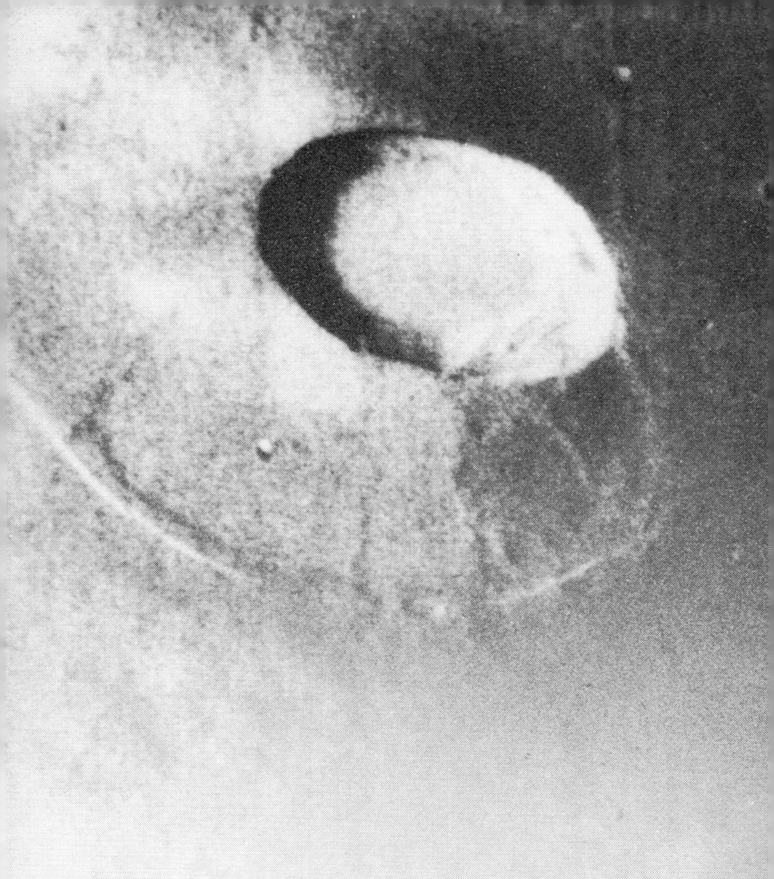

Mars is red because it is rusty. Dark, basaltic material is produced by Martian vulcanism. But why does this basalt rust? (Could it be that Mars once had oxygen and liquid water, which is what makes terrestrial rocks red?)

Global dust storms on Mars are to some extent predictable. They occur each time the planet, in its elliptical orbit, comes closest to the Sun. Fine dust, falling out of the atmosphere in the waning stages of a dust storm, preferentially accumulates in crater interiors. High velocity winds then blow the dust out all in the same direction, producing lovely crater tails, natural weather vanes which tell the meteorologist on Earth which way the Martian winds once blew.

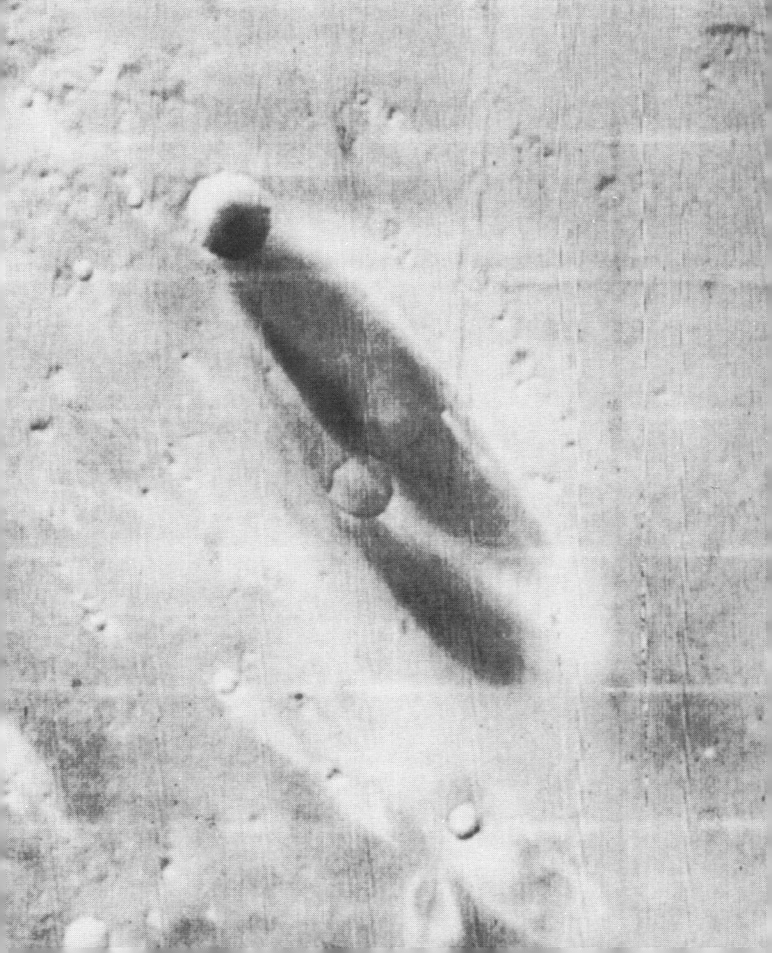

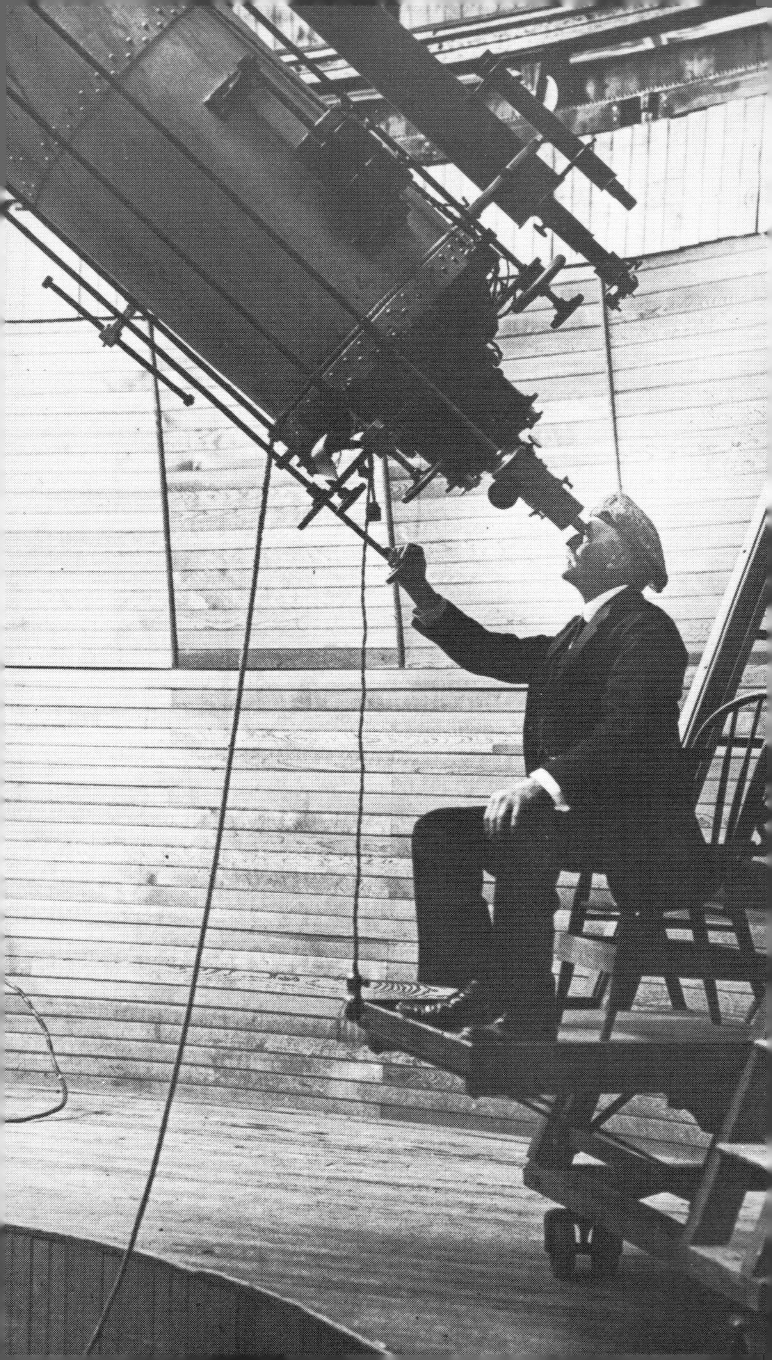

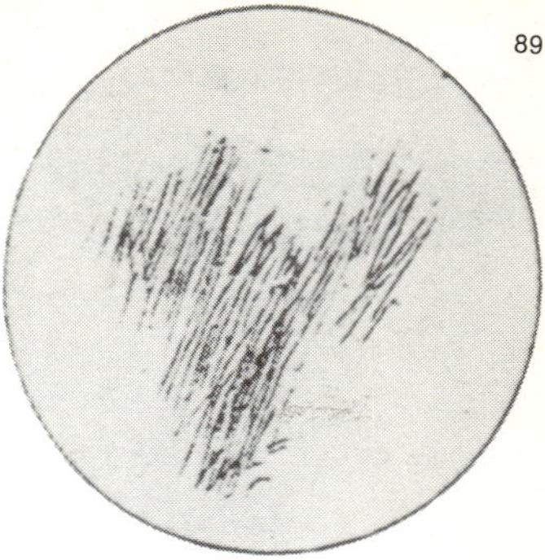

28 novembre 1659

The American astronomer Percival Lowell thought he saw a network of fine, straight, single and double canals criss-crossing Mars, and concluded that they were produced by a race of intelligent beings with a passion for water conservation. But when we examined Mars 75 years later with Mariner 9, we found nothing like the Lowellian canals. One canal corresponds to the great rift valley, Valles Marineris. One might be a ridge and perhaps two or three are the accidental lining up of impact craters. But the hundreds of remaining canals correspond to no feature on the planet whatever. The canals were in the eye of the beholder, and are an object lesson on the dangers of less than scrupulous care on any issue connected with extraterrestrial life.

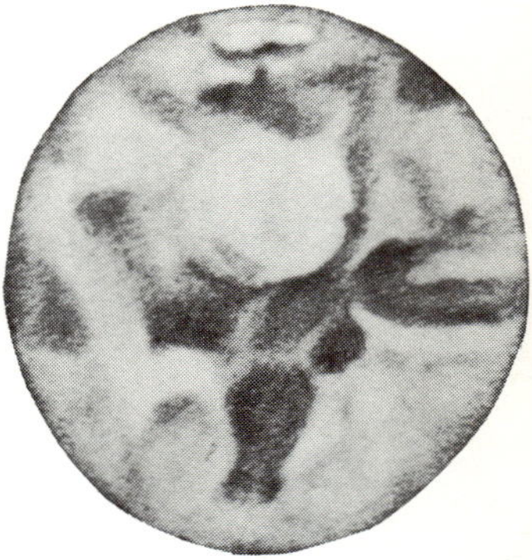

4 septembre 1956

"I am no pilot;
Yet, wert thou as far
As that vast shore
Washt with the furthest sea,
I would adventure
For such merchandise."

—*Romeo and Juliet,* II, i.

THE FLOATING CITY

There is an old story about the Biology One examination in which the students were asked: "Suppose you could take to Mars any of the laboratory equipment used in this course. How would you determine if there were life on Mars?" One famous response: "Ask the inhabitants. Even a negative answer would be significant." The student got an A.

But the problem is probably somewhat more difficult. In 1976 two exquisitely miniaturized automated laboratories called Viking will land on Mars. Mechanical arms will carry soil samples into a series of hoppers within the spacecraft. An elaborate chemistry set will look for organic molecules. Three different biology experiments will search for the metabolism of Martian microbes. In addition, there will be cameras to record any macrobes that saunter by. The data will be radioed to Earth, many tens of millions of miles away.

The prime Viking landing site is Chryse, a Greek word for "the land of gold." Chryse lies at the confluence of several sinuous channels. Perhaps Martian microbes live there, preferentially, because the channels may have been the last places with water before Mars dried up.

Two different mass spectrometers will search for noble gases in the Martian atmosphere. These gases are chemically unreactive and do not freeze even at the cold poles of Mars. Should Mars have had at an earlier time a much denser atmosphere, substantial quantities of argon should still be around today.

A preliminary report of an experiment on board the partially successful Mars 6 descent module of the Soviet Union suggests the discovery in the Martian atmosphere of about 35 per cent of a chemically unreactive gas, probably argon. The details of this experiment have not yet been released; but if the conclusions are valid, they provide important evidence of a denser atmosphere and a more clement environment in the Martian past.

But if some day there is a positive indication of life on Mars, we will have examined two planets and found life on both of them. If the biologies are independent, such a discovery would imply that the origin of life does not require remarkable or special or unlikely conditions. Many people then would be willing to take the grand leap to the view that life is a common occurrence on innumerable planets of other stars in the Galaxy.

Viking may not be a sufficiently fine comb for finding the Martians—if there are any. That might require, for example, an automatic roving vehicle able to examine many different Martian locales.

"Oh, really? What part of Mars?"

Should it ever be decided that samples of Martian soil will be returned for analysis on Earth, the sample will require rigorous containment, and probably sterilization as well. For we know nothing of the nature of conceivable Martian pathogens. Perhaps, because we have evolved differently, the two life forms will be unable to interact; but perhaps, for the same reason, we will have developed no defenses against Martian microbes. Back-contamination of Earth may be a remote contingency; but if it occurs, the consequences are too serious to gamble with.

Insurance premiums are set the same way. It is unlikely that my house will be struck by lightning, but I value my house enough to take appropriate precautions against this unlikely event.

Unmanned biological explorations of Mars will help to narrow the possibilities of back-contamination of Earth.

Mars was a stoplight in the north sky, the only scrap of meat on the night's black bones. And I said: Mars, why be parsimonious? You've got a million tricks stashed in your orbital backhills: chicory suns bobbing in viridian lagoons; quasars dwindling near the speed of light; dumbbell, pinwheel, and impacted galaxies; epileptic nuclei a mile long; vampiric moons; dicotyledon suns; whorling dustbowls of umbilical snow; milky ways that, on the slant, look like freshly-fed pythons.

—Diane Ackerman

We may owe our very existence to climatic changes that on the average amount to only a few degrees. Such changes have brought some species into being and extinguished others. The character of life on our planet has been powerfully influenced by such variations, and it is becoming increasingly clear that the climate is continuing to change today.

We live on a planet on which agricultural technology is responsible for the food of more than a billion people. The crops have not been bred for hardiness against climatic variations. Humans no longer make great migrations in response to climatic change, or at least this is more difficult in a planet controlled by nation-states. It is becoming imperative to understand the causes of climatic variations and to explore the possibility of performing climatic re-engineering of Earth.

Some of the most interesting hints on the nature of climatic change appear to be coming from studies not of Earth but of Mars. Mariner 9 data provide a clear example of the cooling of a planet by the massive injection of dust into its atmosphere. It is also possible that Earth and Mars have undergone synchronous ice ages brought about by variations in the Sun's brightness.

Insights gained from the study of one planet will inevitably aid the study of the others. Comparative planetary climatology appears to be a discipline in the process of birth, with major intellectual interest and practical applications.

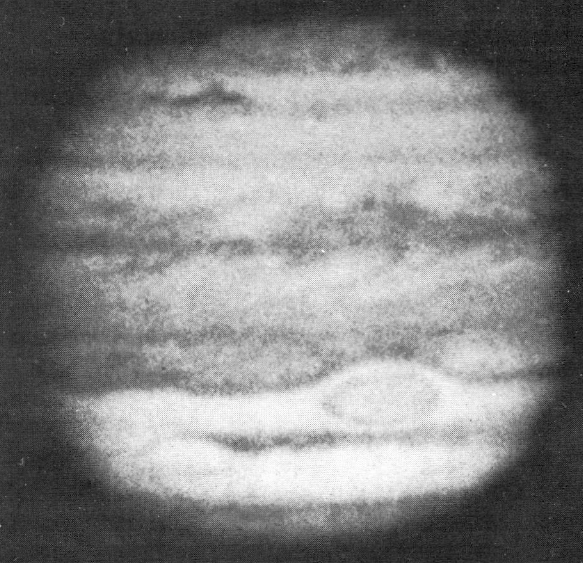

"Dark and deep,
And thick with clouds o'erspread.
Mine eye in vain explored its bottom,
Nor could ought discern."

—Dante Alighieri

The Great Red Spot (GRS) of Jupiter is a "permanent" feature in the Jovian clouds. Half a dozen Earths would fit within it. It is long-lived and it is red. We used also to think it was unique, but Pioneer 10 uncovered a smaller but otherwise perfect replica.

There are three principal explanations of the GRS. In the first, it is a hunk of something, probably frozen hydrogen, floating in the hydrogen-helium atmosphere of Jupiter. But a solid hydrogen raft is not red and may not survive the high pressures of the Jovian atmosphere. In the second, the spot is the top of a tall, completely stagnant tube of atmosphere called a Taylor Column, attached to a small surface irregularity. But there seems no good reason why a stagnant column should be red; its permanence would be puzzling; if it is attached to a surface irregularity, the column would have wound three times around Jupiter in the last century. In addition, clouds are observed moving into and out of the GRS, inconsistent with the idea of a stagnant column.

The remaining possibility, which I have favored, is that the GRS is an enormous storm system. The red coloration is then due to the upwards convection of red material synthesized at depth. We see this red material elsewhere on the planet when there are breaks in the overlying white ammonia cirrus clouds. We see down to reddish places like the North Equatorial Belt; when infrared instruments examine such a region, we find them to be around room temperature.

One of the most interesting possibilities is that reddish organic molecules are made by thunderstorms in the lower aqueous clouds. The meteorology and the organic chemistry of roiling, tempestuous Jupiter are two of the major scientific challenges facing us in the forthcoming exploration of the outer solar system.

"Out, hyperbolical fiend!"

—*Twelfth Night,* IV. ii.

One of the most interesting findings from Comet Kohoutek was the discovery of simple organic molecules, such as hydrogen cyanide and acetonitrile, suggesting that comets share a common chemistry with the gas and dust between the stars.

Comets mostly live half-way between here and the nearest star, and move slowly around the Sun, which itself is only a point of light at that distance. An occasional gravitational tug by a passing star perturbs the motion of the floating iceberg, which then falls into the inner solar system where solar radiation produces a spectacular tail from the vaporizing ice. The tail always points outward from the Sun, whether the comet is approaching or receding from our star.

Comets are denizens of interstellar space, and they may be the first large-scale samples of interstellar matter that we will have to study. Since the interstellar medium is replete with organic molecules, might not the organic chemistry, if not the biology, for the origin of life have come from interstellar space? Molecules critical for biology might have been formed in interstellar space, landed on Earth, and greatly accelerated the origin of life on our planet.

But there are problems: The rate of production of such molecules on the primitive Earth was enormous and the rate of arrival from the interstellar medium very small. Also, such molecules falling on the forming Earth would have been heated to high temperatures, because an Earth hot enough to melt rock was certainly hot enough to fry organic molecules. If such critical molecules did come pouring down from the skies, they would have been destroyed before making any contribution to the origin of life.

It is stunning that organic molecules are abundant in the cold and radiation-filled space between the stars. The relevance of interstellar organic molecules for the evolution of life on Earth is analogical, not substantive. It shows that the chemistry for the origin of life is common. Organic molecules are found in extremely varied and even apparently hostile conditions.

Darwin thought, in one of the earliest reasonable speculations on the origin of life, that there may have been some isolated little pool on the early Earth that by chance had the right combination of salts and organic molecules to produce the first living thing. This kind of thinking—that very special conditions must have been required for the origin of life—has persisted.

However, the rate of production of organic molecules in laboratory simulation experiments is astonishingly high. For example, in our laboratory work at Cornell we use long wave-length ultraviolet light to produce amino acids, the building blocks of proteins, from a simulated primitive atmosphere. We produce amino acids very efficiently. In fact, we produce one amino acid molecule for every 100,000 ultraviolet photons that are absorbed. To translate this efficiency to the primitive Earth, we must allow for the fact that molecules are destroyed as well as synthesized.

We find that the steady-state abundance of amino acids was so high that, if dissolved in the present oceans, they would result in a one per cent solution. Since many other kinds are also produced in such experiments, the primitive oceans were very likely a several per cent solution of organic matter. As Leslie Orgel has noted, this is the approximate composition of Knorr's chicken soup, which is at least alleged to sustain life.

"Thou com'st in such a questionable shape
That I will speak to thee: . . ."

—*Hamlet*, I. iv.

I. S. Shklovskii is one of the leading astrophysicists in the world and, more than any other person, the scientist who made the Soviet Union a leading center of activity in the search for extraterrestrial life. It was my great pleasure to write with him *Intelligent Life in the Universe,* a book that seems to have played a role in arousing attention about this most exciting quest.

Shklovskii is a very human human-being. He combines the critical judgment of the best scientists with a rarely paralleled sense of humor. I was once sitting next to Shklovskii at a scientific meeting in which a speaker seriously proposed that the greatest scientific achievements had been accomplished in periods of maximum solar activity—when there are many sunspots visible on the solar disc and frequent energetic flares spew solar material into space. It was a thinly disguised astrological apologia. The speaker declared that Newton's, Darwin's, and Einstein's greatest works were performed during solar maxima. Then Shklovskii leaned over to me and said in a loud stage whisper, "Yes, but this theory was conceived in a deep solar minimum."

THE SUN.

NEW YORK, FRIDAY MORNING, AUGUST 28, 1835.

GREAT ASTRONOMICAL DISCOVERIES LATELY MADE

BY SIR JOHN HERSCHEL, LL.D. F.R.S., &C.

AT THE CAPE OF GOOD HOPE

FROM SUPPLEMENT TO THE EDINBURGH JOURNAL OF SCIENCE

(*continued from yesterday's Sun*)

As the almost interminable sweep was measured off upon the canvas, we frequently saw long lines of some yellow metal hanging from the crevices of the horizontal strata in wild net work or straight pendant branches. We of course concluded that this was virgin gold and we had no assay-master to prove the contrary. On searching the plain, over which we had observed the woods roving in all the shapes of clouds in the sky, we were again delighted with the discovery of animals. The first observed was a quadruped with an amazingly long neck, head like a sheep, bearing two long Spiral horns, white as ivory, and standing in parallel to each ... that of

walking erect towards a small wood near the base of the eastern precipices. Certainly they were like human beings, for their wings had now disappeared, and their attitude in walking was both direct and dignified. Having observed them at this distance for some minutes we introduced lens, H. which brought them to the apparent proximity of eighty yards: the highest clear magnitude we possessed until the latter end of March, when we effected an improvement in the gas-burners. About half of the first party had passed beyond our canvas; but of all the other we had a perfectly distinct a[nd] deliberate view. They avera[ged] four feet in height, were cove[red] except on the face, with [...] and glossy copper-colored [hair] and had wings compose[d of a] thin membrane, withou[t hair,] lying snugly upon thei[r backs] from the top of the sho[ulders to] the calves of the legs. [...] which was of a yell[ow ...] color, was a slight i[...] upon that of the lar[...] these creatures w[...] engaged in conv[ersation ...] gesticulation, mo[...] the varied actio[n ...] and arms appea[red ...] and emphatic. [...] that they wer[e ...] and althoug[h ...] high an ord[er ...] covered th[...] Bay of Ra[...] capable [...] art and [...]

On August 25, 1835, in the midst of a heat wave and a circulation war, The Sun, a New York daily, printed a startling revelation:

It had been known that the celebrated British astronomer Sir John Herschel, son of Sir William Herschel, discoverer of the planet Uranus, had established in South Africa the world's largest telescope. The Sun announced that Herschel had discovered life on the Moon.

Every day for a week more details were revealed. Among the many organisms reportedly viewed by Herschel were furry bat-like creatures. "Those of the creatures that we saw bathing in the water spread their wings to their full width, waved them as ducks do to shake off the water, and then instantly closed them again in compact form...these creatures were evidently engaged in conversation; their gesticulation . . . appeared impashioned and emphatic . . . they were capable of producing works of art and contrivance."

It was easily calculable then, as now, that nothing smaller than a few hundred feet across could possibly be seen on the Moon from the Earth. No matter how large your telescope, the shimmering of the Earth's atmosphere prevents you from seeing very fine detail. Even if the Moon were covered wing-tip to wing-tip with batmen, they could not be seen with Earth-based telescopes.

It was a hoax, pure and simple and momentarily successful: The Sun's circulation reached an all-time high of 20,000 copies. But the hoax did not last even as long as the round-trip time for the packet boat from New York to Cape Town.

The Sun does not exist today. The only echo of the lunar batman may be the comic book hero of the same name. But the incident shows that a century and a half ago, as now, there were unscrupulous publishers and authors happy to promote sales with whole-cloth inventions about extraterrestrial intelligence.

"No testimony is sufficient to establish a miracle, unless the testimony be of such a kind that its falsehood would be more miraculous than the fact which it endeavors to establish: . . ."

—David Hume

There are well-documented sightings of unidentified flying objects (UFOs), including some by congressmen, pilots, astronauts and policemen. By definition the objects responsible for these sightings are unidentified. Sighting a UFO implies nothing about extraterrestrial visitation—only that something was seen that was not understood. I have no quarrel with those who see unidentified flying objects. It is only when they are identified that I sometimes have misgivings.

As long as people are credulous and soft-minded, and as long as their wishes determine their beliefs, there will be a market for myths and prevarications dressed up in the robes of science. The UFO and ancient-astronaut businesses are businesses. A book or a television program will attract much more attention if it claims that extraterrestrials have visited or are visiting the Earth than if it claims that no valid evidence exists one way or another on this fascinating question.

The remarkable success of such books as Erich von Däniken's *Chariots of the Gods?* has naturally led to a multitude of imitators even more careless with the facts than von Däniken. But the imitation is reciprocal; many of these books were published in their original editions long before von Däniken's.

I was once called by the fledgling author of a UFO book who told me he wished to benefit by the opinions of all serious students of the subject. He assured me that he had an open mind, that he was in the early stages of writing his book and that he had no predisposition to one answer or another. I told him I was very pleased to hear this because the subject was riddled with all sorts of strong tilts. But when I asked him the title of the book, he replied it would refer to human communication with extraterrestrial spacecraft. How, I asked, was such a title consistent with one healthy possibility of UFO origin, namely, that the reported sightings are misapprehended natural objects and have nothing to do with extraterrestrial visitations? Well, he answered somewhat ruefully, the publisher would sign his contract only if a title referring to extraterrestrial communication were written into it. When I replied that this stipulation did not sound like the dispassionate pursuit of truth, he answered that he had clergymen on both sides of his family and therefore could hardly be unethical.

A great fireball almost grazed the American mountain states in 1972—a meteor entered the Earth's atmosphere, exhibited a spectacular trail, and just as suddenly departed. It has been estimated that the fireball came within 36 miles of the Earth's surface, that it measured about 40 feet across, that it weighed about 4000 tons, and that if it had impacted, it would have equalled the yield of the Hiroshima nuclear weapon (the equivalent of about 20,000 tons of TNT). The trajectory corresponded perfectly well to the trajectory of natural objects in our solar system; there was nothing in the sighting to suggest it was an extraterrestrial spacecraft.

This 1972 bolide has relevance for those who believe that many UFO sightings are due to extraterrestrial spacecraft. For what is extremely striking is that hundreds of people observed the event and dozens of them photographed it with still and motion pictures; as far as can be determined, all of the visual accounts are mutually consistent. They correspond closely to the same trajectory, the same relative brightness, the same color, and so on. Given this massive body of data, so beautifully consistent, it is possible to use the methods of modern physics and astronomy and deduce the nature of the fireball. If on the other hand, only one person had seen it, or if there were only one fuzzy photograph, it would remain forever in the category of the unknown.

The 1972 bolide is not nearly as unusual an object as many described in the UFO literature. How is it that in the UFO literature there are no descriptions of comparable consistency by hundreds of people and no similarly detailed motion picture films? Could it be that extraterrestrial visitation is deducible only in those cases where the data are marginal or of low quality or uncorroborated or otherwise suspect?

I don't know why UFOs are never sighted over large cities by hordes of people. But it is consistent with the idea that there are no space vehicles from elsewhere in our skies. I suppose it is also consistent with the idea that space vehicles from elsewhere avoid large cities. However, the primary argument against recent extraterrestrial visitation is the absence of evidence.

Take leprechauns. Suppose there are frequent reports of leprechauns. Because I myself am emotionally predisposed in favor of leprechauns, I would want to check the evidence especially carefully. Suppose I find that 500 picnickers independently saw a green blur in the forest. Terrific. But so what? This is evidence only for a green blur. Maybe it was a fast hummingbird. Such cases are reliable but not particularly interesting.

Now suppose that someone reports: "I was walking through the forest and came upon a convention of 7000 leprechauns. We talked for a while and I was taken down into their hole in the ground and shown pots of gold and feathered green hats." I will reply, "Fabulous! Who else went along?" And he will say,

"Nobody," or "My fishing partner." This is a case that is interesting but unreliable. In a case of such importance, the uncorroborated testimony of one or two people is almost worthless. What I want is for the 500 picnickers to come upon the 7000 leprechauns . . . or vice versa.

The situation is the same with UFOs. The reliable cases are uninteresting and the interesting cases are unreliable. Unfortunately, there are no cases that are both reliable and interesting.

I once noticed a young woman reading a typical page in *Chariots of the Gods?* in which the author grossly misquoted someone; in this case, me. I asked her if she were enjoying the book, and she said, "Yes, immensely," with a kind of pious fervor in her eyes. I said that I had heard that some of the book was inaccurate, but she quickly assured me that *that* was quite impossible. Why? "Because they wouldn't let him publish it if it weren't true."

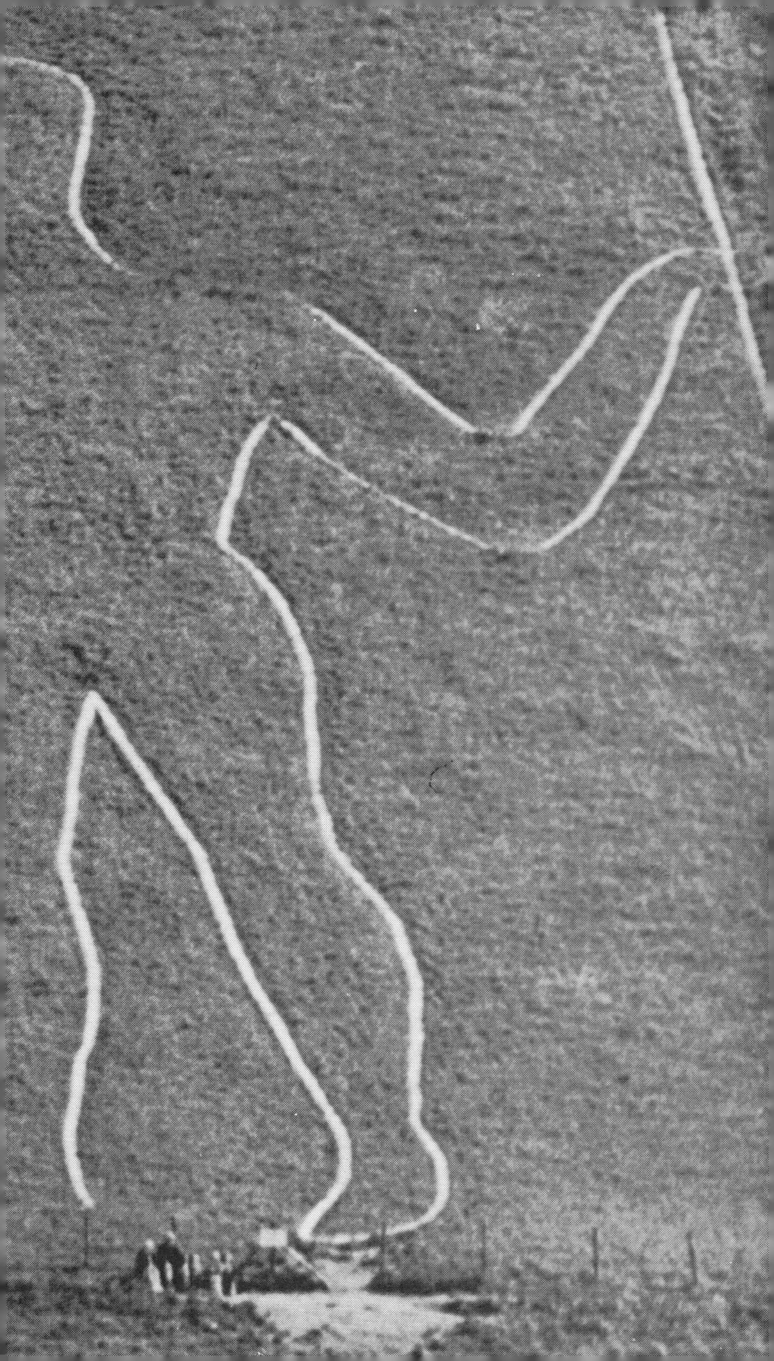

The French archaeologist Henri Lhote discovered in North Africa a set of beautiful wall paintings of great antiquity called the Tassili frescoes. Lhote and his colleagues had called one of these paintings, perhaps jocularly, "The Man from Mars." I. S. Shklovskii and I discussed this painting in *Intelligent Life in the Universe* and found no evidence for any extraterrestrial connection. Subsequently, von Däniken and his imitators naturally claimed that the fresco shows an extraterrestrial being in a spacesuit. It is remarkable how similar the extraterrestrial is to us, except perhaps for having only a single eye.

I was very struck when my son Nicholas brought home from nursery school a cutting, the outline of his body as he lay down on a piece of paper. He later drew in his 3½-year-old features. The similarity to the Tassili fresco is certainly as close as the similarity of a space-suited astronaut to the fresco.

Since many cave paintings include the tracings of the hands of the artists, I wonder if it is not more likely that a full-body tracing or a shadow projection occurred in North Africa than that an extraterrestrial visitor was commemorated on a wall.

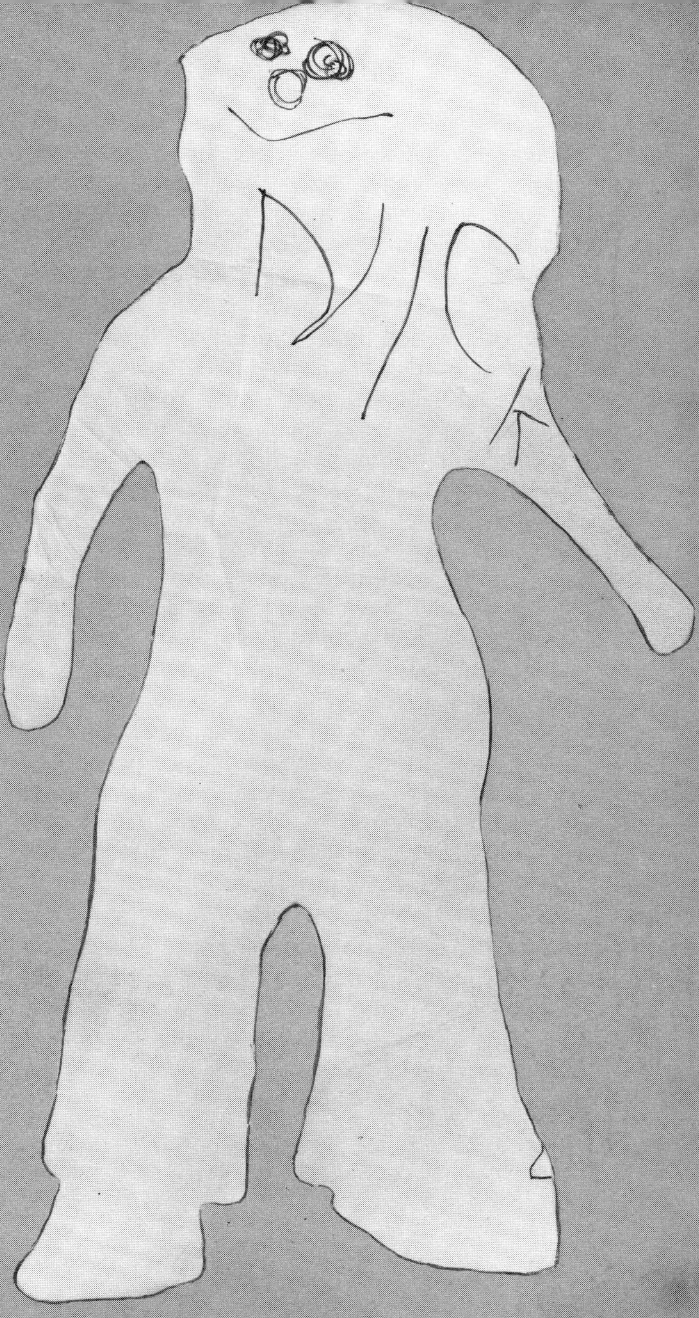
DICK

When there was no prospect of getting to the planets, many mystics were happy to tell us what the planets were like. Now that we have performed an initial reconnaissance of the planets, these accounts are found to have little resemblance to the true planetary environments, and contemporary mystics are found to be slightly more hesitant.

Recently, two courageous American mystics made an "astral projection" trip to Jupiter, describing the nature of the planet prior to the arrival of Pioneer 10. I was asked to examine the accuracy of their account. If their reports had been submitted in my elementary astronomy course, they would have received grades of "D." Their reports were not better than what can be extracted from the worst popularizations of planetary astronomy; they were filled with the most obvious misunderstandings both about Jupiter and about Pioneer 10.

As just one example, Pioneer 10 was reported shaped like a bullet—perhaps the average man's most naive picture of an interplanetary spacecraft. In fact, interplanetary spacecraft are not streamlined because they do not enter planetary atmospheres. They may have, and Pioneer 10 does have, all sorts of odd and convenient shapes because there is negligible atmospheric resistance in the medium through which they travel.

There is no evidence that any mystic has done better in guessing the nature of the planets than he could do without his mystical powers but with the ability to read the better elementary astronomy books.

Drawing by Chas. Addams; © 1962 The New Yorker Magazine, Inc.

Astrology has been with us for at least 5000 years. This does not imply its validity, but only its resonance with what we need to believe.

To the best of my knowledge, there has never been published in a recognized and refereed scientific journal a paper arguing the validity of astrology. If the evidence for astrology is plentiful and readily available, why is it not published in the scientific literature?

There is some evidence that people in certain specific occupations tend, more often than by chance, to be born at certain months of the year, or, as we might say, under certain astrological signs. But young humans might be very powerfully influenced either by pre-partum or by immediately post-partum climate. In the temperate zones, where most surveys have been made, January's weather is very different from July's.

No mechanism has been proposed for astrology. The alleged astrological influence cannot be propagated by light because most people are born in closed buildings. Neither can it be propagated by gravity, because the gravitational influence of the obstetrician is much greater than, say, that of Mars. Mars weighs more, but the obstetrician is a lot closer.

Astrology takes no note of x-ray, radio, gamma ray and infrared sources and many solar system objects discovered in the last few thousand years. Its practitioners often employ incorrect tables of planetary positions and make no allowance for atmospheric refraction.

The popularity of astrology is an unhappy commentary on the lack of toughmindedness and the dearth of open and critical thinking in our society.

Scientists are the worst people imaginable to check out frauds. Scientists are used to ferreting out the secrets of nature. But nature fights fair. Frauds do not.

To check out many sorts of frauds we need magicians, not scientists. Some of the leading British physicists were presidents and other officers of the Psychical Research Society; the spirit world was big in England at the turn of the century. But when the magician Harry Houdini examined these psychical matters, frauds were exposed left and right: He reached under the table and found the wooden thumper; he turned on the light and found the assistant holding the cheesecloth; and so on.

It is fascinating that some of the most widely publicized potential frauds in recent times are delighted to have large audiences of scientists check them out, but blanch at the scrutiny of a single professional magician. The magicians are the ones with a working knowledge of human gullibility.

"The chess-board is the world; the pieces are the phenomena of the universe; the rules of the game are what we call the laws of Nature. The player on the other side is hidden from us. We know that his play is always fair, just, and patient. But also we know, to our cost, that he never overlooks a mistake, or makes the smallest allowance for ignorance."

—T. H. Huxley

Occam's Razor is not perfect; but it has proved to be an extremely useful guide when we are otherwise ignorant. Occam's Razor is called a razor because it slices complex problems exceeding fine. It suggests that, when faced with two equally workable explanations of the same data, we should choose the simpler.

For example: There are three independent chronometers, employing radioactive decay, that give mutually consistent results on the ages of old rocks. But we can hypothesize that a year ago November a mischievous deity created the Earth with false radioactive clues embedded in it. No one could ever disprove such a possibility. All our memories, all our written history, could have been created at the same moment. But the simpler hypothesis is that the chronometers indicate what they seem to indicate: That the Earth formed 4.6-billion years ago; that the major mountain ranges are millions to billions of years old; and so on.

The mythological apologetics of Immanuel Velikovsky imagine that the major mountain ranges on the Earth and the major craters on the Moon were created no more than 3500 years ago. Since the radioactive chronometers for mountains and returned lunar samples give ages thousands to millions of times older, Velikovsky airily dismisses the radioactive dating methods despite their great mutual consistency. He and his followers are forced to stress discrepancies of tens of per cent, when what they need are non-existent discrepancies of millions to billions of per cent. It is much simpler to conclude that Velikovsky's interpretation of ancient myths is in error.

Many similar examples, where Occam's Razor could be employed to advantage, can be culled from the literature of folk or border-line science: It is not excluded that a germ of truth or the glimmering of a useful insight might reside in one of these improbable hypotheses. But the vast majority do not cut.

"Very well then, hands up all those who propose to become birds."

Man was neither premeditated nor a mistake. The clear lesson of evolution is that a long series of fortuitous changes resulted in the origin of man and all the other beasts.

Somewhere else: another set of changes, another set of beasts.

There is powerful biochemical and morphological evidence of the deep connectedness of humans with all the other organisms on Earth. If we came from elsewhere, so did all the other organisms, great and humble, complex and simple.

But could life on Earth have derived from organisms from elsewhere? If so, it must have occurred four-billion-or-so years ago and have involved organisms no more complex than the bacteria and the blue-green algae. One suggestion is that such ancestral bugs were driven across interstellar space by radiation pressure, the outward impulse that starlight gives to particles in a certain size range. Only very special combinations of donor and acceptor star systems permit such a traverse. The travel times would not be huge—only millions of years.

But there is the serious problem of radiation damage. To leave, the organism would somehow have to get off its planet into interplanetary space, and then be ejected from its solar system by the pressure of starlight. Small organisms are very sensitive to radiation damage. Generally, they cannot be shielded without making them too big to be ejected. A terrestrial alga or bacterium placed just above the Earth's atmosphere would be killed by solar ultraviolet light in about one second.

Another suggestion is that the Earth was accidentally seeded four-billion-or-so years ago by intelligent beings from another star. We can imagine a picnic on the virgin Earth in which the microörganisms in the cookie crumbs are the ancestors of us all.

It has even been suggested that an interstellar spacecraft was intentionally dispatched to seed the Earth—something like the arrival of Superman from Krypton, but with a longer time scale. An interstellar Johnny Starseed is perhaps a better image.

Such ideas are by their very nature impossible to disprove. The need for such hypotheses is not compelling. We understand in sufficient detail most of the steps that lead to the indigenous origin of life on a planet. The molecules of life were made readily on the primitive Earth. There seems no need to explain its origins by artificial insemination. Such seeding hypotheses postpone rather than solve the problem of the origin of life.

Amino acids are easy to make. But no one has yet started with a mixture of gases and wound up with a small beast crawling out of the test tube. The Earth experimented for billions of years before achieving such a result; we have only been trying for twenty.

Exobiology, a term coined by the Stanford University geneticist and Nobel laureate Joshua Lederberg, is the study of life elsewhere. Since we have not yet found life elsewhere, exobiology is a speculative subject. It includes the study of pre-biological syntheses of organic molecules that lead to the origin of life; the pre-technological evolution of life, if any, on planets of our own solar system; and conceivable technological societies on planets of other stars.

"Have you ever wondered, Muldoon, if life as we know it exists in other bars?"

"And we come in four basic colors—red, black, white, and yellow."

Apes are not evolving into men today in part because it requires a very special set of environmental circumstances. Any organism that offers serious competition is eliminated by man. This is the clear lesson of archaeology and history.

Human history is full of internecine mass murder committed on the grounds of slight differences in form, dress or skin color. This xenophobia, this fear of the unfamiliar, may have had selective value at one time. It no longer does.

If we are some day to feel comfortable in the presence of beings who have evolved independently and on other worlds, we must transcend the feeling "If it's different, step on it."

"All right, all right, class—now that we've had our little laugh . . ."

"It is the business of the future to be dangerous."
—A. N. Whitehead

Predicting the future is tough.

LA MODE

There are no detailed predictions of our time that have proved even roughly accurate. Our epoch is unpredictable because it is simultaneously complex and changing rapidly. This seems also to be the reason for the madness of our times. There is no moment in the history of mankind when so many changes in so many different areas—social, political, economic, scientific, technological, sexual and educational—have occurred. They are happening too fast for too many people. Madness is one way of coping.

Every adult alive today can recall enormous changes that have already happened within his own lifetime. A generation is a reasonable time scale for significant change. Shorter periods are not. Only intelligent and sensitive planning, with full attention to the uses and misuses of technology, can prevent the future from being excessively dangerous.

I got turned on to stars and planets before I knew there was such a thing as science fiction. When I was maybe ten years old, I came upon Burroughs, Wells and Verne and thought them marvelous. At eleven, I discovered Astounding Science Fiction at a newsstand and was exhilarated and delighted with its cosmic perspective.

Since then, I have learned more science than I knew when I was eleven, and my literary tastes have perhaps improved. I am not the same person I was when science fiction carried me on soaring flights of the imagination.

When I return to a science-fiction novel that I last read when I was a boy (and still remember with the greatest pleasure), it often seems to have deteriorated strangely over the years. What has really happened, I think, is my slow discovery that science is far more exciting than science fiction, far more intricate, far more subtle—and science has the additional virtue of being true.

Science fiction today still plays a major role in interesting young people in scientific endeavors, or at least priming them to pay attention to scientific things. This is an important social good.

There is today a clear tendency away from physical and biological science fiction, which may be a defensive maneuver by science-fiction writers who are finding established literary conventions about Mars and Venus and Jupiter crashing down at their feet. But the trend to social and psychological science fiction is also a social good. If we are to solve the difficult problems of our future, it is essential to get bright youngsters interested early in thinking about the evolution of societies.

Science, mathematics and technology dominate our lives and are the hallmark of our civilization. They represent the most essential differences between our world today and that of our Pleistocene ancestors.

More than one-third of the inhabitants of the planet Earth owe their food supplies to agricultural technology. Most of the comforts and conveniences, the leisure time, and many of the sources of intellectual excitement and artistic endeavor in our society derive from science and technology.

But these same forces can be used for evil as for good. Through improved medical and public-health practices, the lifetime of human beings on the planet Earth has almost doubled since Roman times, and the population has accordingly increased geometrically. But through misapplication of the powers of technology, we have created industrial pollution, the exhaustion of mineral and fossil-fuel resources, and perhaps deleterious changes in the climate of our planet.

Nuclear energy and rocket motors are the supreme examples of the choices technology sets for us: To harness the power of the Sun for abundant energy —or to allow destruction of our civilization in a nuclear holocaust; to carry nuclear destruction anywhere in the world—or to transport us to the planets and the stars. There is no turning back from technology. The problem is how to use it well.

Of the 435 members of Congress there are only two or three with professional training in science. There are no Senators with such backgrounds. In the Nixon Administration there were no scientists with ready access to the highest level of government. The situation is sadly similar in other nations. The impoverished understanding of science at the "top" unfortunately but understandably reflects the situation in the public at large. This is madness.

Exacta (6-5) paid $72.40.
EIGHTH—$9,000, allow., 3YO, 6 f.
ver Florin ....(Solomone) 7.00 3.60 3.40
uddy York ....(Blum) ... 3.40 2.60
alman ....(Perret) 3.80
Time 1:10 1/5. David's Pinecone, Flash t, Oritani, The Grok also ran.
NINTH—$6,000, cl, 3YO, 1 1/16 m.
causuc ....(Solomone) 18.40 14.20 5.40
lf Fisherman ...(Garcia) ... 7.00 4.60
No Vivie ....(Kallai) 5.40
Time 1:46 1/5. One Good Cent, Taxi For axie, Delayed Orbit, Sir Pardy, Doctor's ld, Hitching Post, Motto also ran.
$3 Trifecta (9-6-7) paid $3,303.30.
Attendance, 17,382. Handle, $2,131,211.
Weather clear, track fast.

Monmouth Entries

OCEANPORT, N.J.

By The Associated Press

Horses listed in order of post positions.

FIRST—$4,300, cl., mdns, 3 and 4YO, 6f.

	Prob. Odds		Prob. Odds
n Me Up, 117	12-1	Tribal Miss, 117	4-1
otsieWootsie,115	8-1	Sly Jay, 117	8-1
's Vicky, *113	12-1	QunlyMonrch,117	12-1
dy Zoe, *106	10-1	Odille, *107	20-1
shver Girl, 117	6-1	Ready Chris,*108	15-1
vely Belle, 117	7-2	Ethel Farkle, *112	3-1

SECOND—$4,500, cl., mdns., and 3 and YO, 6f.

rward Snap,113	8-1	RunningBuck,*112	7-2
ldivia, 113	12-1	Ragging Wrr, *105	8-1
essedCorrel,120	15-1	FearlessDevil,117	10-1
aginaryLn,*109	20-1	Martin Vale, 112	12-1
ince Vigor, 117	15-1	U.CuteBuperU,117	15-1
ieEmperor, 120	4-1	Dsh n' Finish,117	20-1
dayPainter,117	10-1	Speedy Smith, 117	5-2
d Notion 116	6-1	Regal Land, 117	3-1
rrcneWtrs*112	12-1	Bold n' Brash, 115	4-1

THIRD—$5,500, cl., 3YO and up, 6f.

appy Lane, 114	10-1	Tooth Maker, *114	6-1
Big Devil, *108	5-2	Young Eagle,115	15-1
nionSoldier,*110	5-2	Bold Beacon, 115	8-1
od Skate, 122	4-1	Big Ken, *114	4-1
een Belt, 119	7-2	Tootieboy, 115	15-1
nChicholl,115	12-1		

u-Coupled.

FOURTH—$6,000, cl., 3YO and up, 1m. urf).

Wmn'sScrt *111	7-2	Atlantic Br'ze 113	3-1
ntilly 116	12-1	Prnc's Anita *102	10-1
ynesian ...116	15-1	Made It 112	6-1
lShrpNail *109	8-1	Mechtilde ..*107	12-1
gusta J. ..120	4-1		

FIFTH—$5,500, cl., 2YO., 5½f.

wh'se Chck 117	6-1	Polar Jo 120	7-2
t'sAffrmatv 115	4-1	Flaring Red .120	5-2
t Stuff Too 117	3-1	Ambit'ninrest 117	10-1
n Toxic ...115	12-1	Just Our Jill 114	12-1
wsh Princs 114	10-1		

World Team Tennis

(Includes matches of Aug. 11)

AT MANHASSET BAY

COW BAY Y.R.A.

Etchells-22 Class (17)—1, Charl Hagen; 2, Howard Seymour; 3, B land and Bud Preston.
Soling Class (6)—1, Walter E. B Patricia Wallmuller; 3, Ronald Gou
Shields Class (8)—1, James C. 2, Karl Maier; 3, Fred Poor.
Knickerbocker Class (7)—1, Jessie berg; 2, Leonard Friedman; 3, L Kamisher.
Lightning Class (2)—1, No. 9912; 2 Ondrey.
Ensign Class (10)—1, Stephen Ru Dr. L. Friedman; 3, William Siener
MBOD Class (6)—1, Bob Kelley; Coverly; 3, Brakeman and Mall
Laser Class (8)—1, Paul Gaddis; Hagedorn; 3, L. Ingersol.
Thistle Class (5)—1, Peter Dobson; Goetz; 3, Herb and Peter Schmitz.
Fireball Class (2)—1, Chris Dani P. Mitropolis.
Blue Jay Class (3)—1, Chris Sc 2, Pedro Lorson; 3, Timmy Christ.
Rhodes-19 Class (10)—1, No. 548; Mogensen; 3, Tenney and Bennet
Sunfish Class (7)—1, Cecil Akre; Barnard; 3, David Akre.
Day Sailer Class (3)—1, Wright 2, Jack O'Donnell; 3, Tom Callen.
Multihull Class (6)—1, Warren 2, Cornell Toth; 3, Jamerson H Keith Thomson; 5, Gene Matthews

AT NARRASKETUCK Y

GREAT SOUTH BAY Y.R.A.

Multihull Class (4)—1, Bill Swartw Symon O'Dea. Series winner—Sw
Flying Scot Class (4)—1, Jerry Sach Stan Cole; 3, Jack Harris. Series w Sachnoff.
Narrasketuck Class (6)—1, Stanley B Jim Grover; 3, Brant Davidson. winner—Buys.
Lightning Class (11)—1, Frank Atkin Alan Pearlman; 3, Frank Marinaccio winner—Atkinson.
420 Class (5)—1, Neil Gargiulo; 2, Knapp; 3, Andrew Goodwin. Series —Gargiulo.
Comet Class (8)—1, Lou Guidone; McCann; 3, Matt Bernius. Series w McCann.
Handicap B Class (8)—1, T. Cohill, C 2, Sailfish No. 9970; 3, Phil Packa dinghy. Series winner—Cohill.
Mariner Class (8)—1, Herb Buerger; Powell; 3, Helmuth Meyoefer Serie ner—Meyhoefer.
Sunfish Open Class (6)—1, Randy Pf 2, Paul Harding; 3, Ren Lueder. winner—Harding.
Sunfish Secondary Class (12)—1, No. 2, Ray Lyon; 3, David Jones. Series —Jones.

Florida 32, Chicago 23

Virtually every American newspaper has a daily astrology column. How many have even a weekly science column?

It is argued that scientific information is too technical. But newspapers routinely publish detailed technical information on baseball, including such vital information as AB, SO, HBP, SLAVG, ERA, W-L and SB. The financial pages run hundreds of column inches of more refined obscurity.

Considering the enormous importance of science, should there not be a science page that gives simple and accurate reporting with some details? The presence of such details in the daily newspaper and in prime-time television programming—with the clear implication that at least some people enjoy them—can provide a major stimulus for the young. It would demonstrate that science and mathematics are acceptable activities not irredeemably difficult, and that some people think they are fun.

There is another aspect of science, one that is infrequently described except among the practitioners themselves: Science as a supreme art form. The creative endeavor in science carries the same emotional exhilaration as the painting of a great work of art or the writing of an epic poem. In science we have a glimmering of the intricate elegance, awesomeness and monumental beauty of nature. It is devastating how well the parts fit together: How the atoms synthesized in the interiors of hot red giant stars are the same everywhere in the Universe; how they combine to form molecules in the same way everywhere in the Universe; how the nature of human beings is powerfully prefigured by the nature of molecules called nucleic acids; how the change in a single small molecule determines the difference between mental health and manic depressive psychosis; between malaria and sickle cell anemia; between life and slow death by phenoketonurea; how a difference in molecules too small to be seen determines the difference between a man and a bacterium; how the same carbon atom joined to its fellows in different ways spells the difference between charcoal and diamond.

For the survival of our civilization and our species, as well as for what philosophers since Socrates have recognized as the most profound of human pleasures, I believe there is an urgent need for better, serious and more widespread popularization and understanding of science.

Oct.	150.00	150.00	150.00	150.00	160.00
Dec.	155.00	155.00	154.20	154.20	164.20
Jan.	157.00	157.00	156.00	156.50	166.00
Mar.	158.00	159.00	158.00	158.50	168.00
May	162.00	162.00	159.20	160.00	169.50
July	162.00	162.00	160.30	161.50	170.50

CHICAGO CASH GRAINS

Wheat No. 2 soft red 4.25½n.
Corn No. 2 yellow 3.43n.
Oats No. 2 extra heavy white 1.68¼n.
Soybeans No. 1 yellow 7.69½n; Chicago high rate.
Barley malting 3.00-3.50n; feed 1.85-2.25n.

POTATOES

	High	Low	Close	Prev. Close
Nov.	4.67	4.36	4.67	4.50
Mar.	5.25	4.97	5.25	5.17
Apr.	5.57	5.25	5.57	5.52
May	6.78	6.37	6.73	6.59

Sales, 3,461 contracts.

99 PLATINUM

New York Mercantile Exchange

Oct.	185.90	182.00	185.80	185.50
Jan.	192.80	188.50	192.50	192.00
Apr.	198.20	194.00	197.70a	197.50
June	202.00	200.00	202.30b	202.50
Oct.	207.00	204.50	207.00	207.5[illegible]
Jan.	211.50	211.50	211.50	211.30

Sales, 527 contracts.

PALLADIUM

Sept.	143.00			142.00

Closings—Dec., 146.00/147.50; March '75, 148.00/152.00. Sales, 2 contracts.

U.S. Silver Coins (in dollars)

Oct.	3,052	2,950	2,955	3,080
Jan.	3,154	3,074	3,065a	3,190
Apr.	3,235	3,200	3,165	3,295
July	3,287	3,280	3,255a	3,390
Oct.	3,405	3,405	3,340a	3,465

Closing—Jan. '75, 3,415/3,420. Sales, 118 contracts.

SILVER

Comm[illegible]xchange, Inc

N.L.	—	.09
N.L.	—	.11
N.L.	—	.11
N.L.	—	.08
4.25	—	.03
2.97	—	.03
5.69	—	.03
5.36	—	.03
N.L.	—	.07
N.L.	—	.03
N.L.	—	.09
8.13	—	.04
10.99	—	.05
.....	—	.02
.....	—	.05
.....	—	.01
.....	—	.02
.....	—	.04
.....	+	.06
16.55	—	.15

Wall St G	[illegible]	[illegible]	—	.04
Wash M	[illegible]	10.63	—	.09
Weing Eq	7.53	N.L.	—	.07
Wellingtn Group:				
Explr	16.41	17.93	—	.02
Ivest	6.05	6.61	—	.08
Morg	8.37	9.15	—	.10
Trust	8.08	8.83	—	.03
Wellsl	9.96	10.99	—	.04
Wellfn	8.56	9.36	—	.06
West B	8.87	9.69	.....	
Windr	6.00	6.56	—	.06
West Ind	2.04	2.24	.....	
Westf Grt	5.51	5.99	—	.07
Wisc Fd	4.30	4.70	—	.05
Ziegler	7.94	8.68	—	.08

N.L. No load (sales charge)

N.L. No load (sales charge)

Stocks not quoted in today's list were unavailable from the NASD.

Unless otherwise noted, rates of dividends in the foregoing

Mar.	56.10	56.10	58.10	57.00

Sales, 1,411 contracts.

FROZEN PORK BELLIES

Aug.	53.25	52.20	52.85/.70t	53.55
Feb.	56.20	56.20	56.20a	57.70
Mar.	56.10	55.92	55.92a	57.42
May	56.85	56.85	56.85a	58.35
July	57.00	56.75	56.75a	58.25

Sales, 365 contracts.

LIVE BEEF CATTLE

Aug.	48.55	47.05	47.75t	47.55
Oct.	48.95	47.25	47.35/.50t	48.12
Dec.	47.40	45.75	45.75a	46.75
Feb.	46.40	44.95	44.95a	45.95
Apr.	46.40	44.97	44.97t	45.97
June	46.95	45.50	45.55/.50t	46.50
Aug.	46.75	45.52	45.60a	46.52

Sales, 12,453 contracts.

LIVE HOGS

Aug.	38.85	37.80	38.60/.50t	37.90
Oct.	37.40	35.60	36.75/.90t	37.05
Dec.	40.10	38.95	39.20/.10t	40.45
Feb.	42.30	41.70	41.70a	43.20
Apr.	42.00	40.80	40.80/.85	42.25
June	44.40	43.30	43.30t	44.80
July	45.00	44.20	44.20a	45.70

Closing—Aug. '75, 42.00n. Sales, 4,240 contracts.

LUMBER

Chicago Mercantile Exchange

Sept.	132.50	130.10	131.10/.00t	132.00
Nov.	128.00	126.10	127.50t	128.0[illegible]
Jan.	128.50	126.90	127.30t	127.5[illegible]
Mar.	131.00	129.00	129.50t	130.0[illegible]
May	132.00	131.00	131.00t	131.0[illegible]

Sales, 349 contracts.

WOOL FUTURES

No reported no sales.

ICED BROILERS

Chicago Board of Trade

Aug.	36.75	36.05	36.25t	36.5[illegible]
Sept.	37.85	36.80	37.6[illegible]	37.4[illegible]
Nov.		38.45	3[illegible]	

				9.00
Jan.	[illegible]	[illegible]	—8	9.0[illegible]
Jan. 2, 1975	8.87	8.55	—6	8.97
Jan. 9	8.87	8.55	—6	8.99
Jan. 14	8.91	8.59	—6	9.04
Jan. 16	8.91	8.59	—6	9.04
Jan. 30	8.95	8.67	—8	9.1[illegible]
Feb. 6	8.83	8.75	—10	9.[illegible]7
Feb. 11	8.77	8.37	—7	8.8[illegible]
Mar. 11	8.79	8.39	—3	8.89
Mar. 31	8.94	8.85	—12	[illegible].43
April 8	8.[illegible]9	8.31	—8	8.82
May 6	8.61	8.33	—10	8.[illegible]
June 3	8.62	8.34	—10	8.9[illegible]
July 1	8.62	8.36	—10	8.96
July 29	8.44	8.36	—10	9.0[illegible]

FEDERAL INTERMEDIATE

Millions outstanding.

593	Sept., '74	8.60	99.25	99.29	9.72
799	Oct., '74	7.95	99.18*	99.22	9.84
754	Nov., '74	8.00	99.12	99.16	9.86
785	Dec., '74	7.15	98.91	99.4	9.92
240	Jan., '75	6.05	98.14	98.22	9.51
608	Jan., '75	8.15	99.8	99.12	9.64
674	Feb., '74	8.80	99.15	99.19	9.46
796	Mar., '75	9.00	99.17	99.21	9.48

An
Ana
Ani
Arp
Ass
Atl
Aut
Aut
Bal
Bar
Bay
Bec
Beg
Ber
Bev
B'g
Bill
Bra
Bre
Bre
Brig
Brn
Bris
Bris
Bru
Bur
CP

"Perhaps the most majestic feature of our whole existence is that while our intelligences are powerful enough to penetrate deeply into the evolution of this quite incredible Universe, we still have not the smallest clue to our own fate."

—Sir Fred Hoyle

The power of sex is astonishing: Toms will spend heroic nights caterwauling; stags will fight till exhaustion; salmon will swim upstream in unparalleled displays of courage. One has only to examine American motion pictures and advertisements (a little introspection should also serve) to see how powerful a drive sex is in human life. Sex has evolved with such relentless power because organisms are generally too dumb to reproduce upon philosophical retrospection alone. The point of sex is to reassort the genetic material and to aid the continuing self-replication of a long double helical molecule called DNA. Our lives are significantly determined by the need to provide convenient reproduction of a molecule few of us have ever seen.

Are we then mere DNA containers briefly employed, paid in orgasmic coin, to transmit the genetic information? I would not say "mere." To be the temporary repository of a four-billion-year-old molecular tradition, threading back to the dawn of the Earth and the birth and death of the stars, is not "mere."

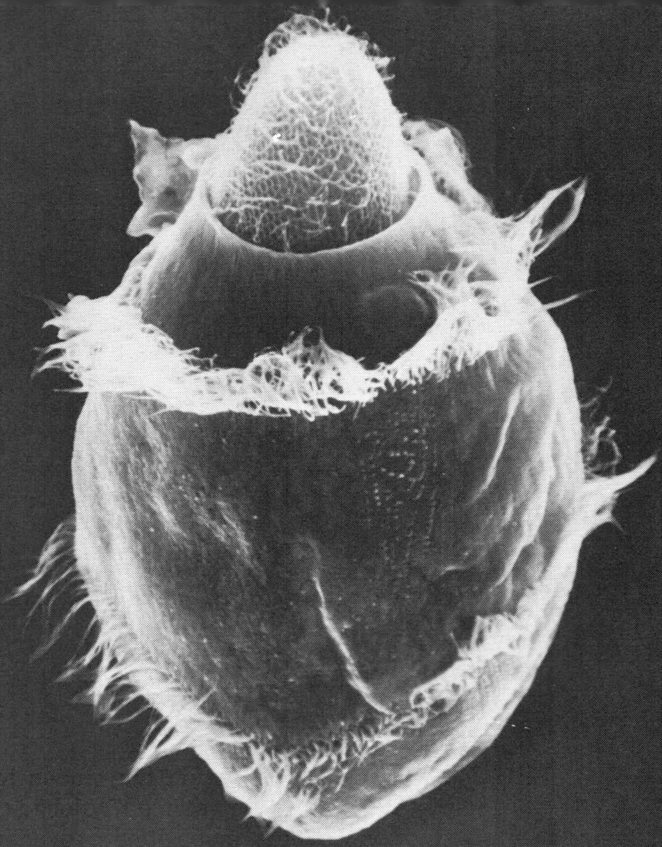

The teachings of modern biology are clear. Human beings may be—if we do not destroy ourselves—an extraordinary evolutionary development, remarkable in our self-awareness and intelligence. But we are fundamentally tied by the deepest bonds of kinship to all the other beasts and vegetables that swim in the sea or fly in the air or grow from the ground or walk upon the Earth or brachiate from tree to tree.

"I always felt that man is a stranger on this planet, a total stranger. I always played with the fancy: maybe a contagion from outer space is the seed of man. Hence, our preoccupation with heaven—with the sky, with the stars, the gods—somewhere out there in outer space. It is a kind of homing impulse. We are drawn to where we come from."

—Eric Hoffer

"Now entertain conjecture of a time
When creeping murmur and the poring dark
Fills the wide vessel of the universe."

—*Henry V,* IV. Chorus, 1.

Often when I give popular talks—and especially when I touch on cosmological questions—I am asked if I believe in God. My usual reply is to ask the questioner what he means by God, for I have found that the widest range of mutually inconsistent ideas is covered by that well-known three-letter word.

A typical response is, "Well, you know, a force more powerful than us that fills the Universe." There certainly are such forces. Gravity is one of them. But not even the most dedicated Spinozan will identify gravity with God, so far as I know.

Most questioners have cosmological problems in mind when they ask about God: How could the Universe have begun if there were no Aristotelian prime-mover? If the Universe was made by some Being, where did that Being come from? And if the Being is eternal, have we solved the problem, or merely postponed facing it by one step? An alternative possibility that removes the infinite regression of causes is this: The Universe itself is eternal; never having been made, it requires no Maker.

I find that people fall about half and half into two different camps of mind-boggle. One camp is boggled by a Universe with an infinite past, although such people are willing enough to admit the prospect of an infinite future for the Universe. This may be because humans and other beings arise at a finite point in time. The other camp is equally boggled by a Universe that is made from nothing.

It is remarkable that such cosmological questions are becoming accessible to experimental inquiry, and in the next few decades we may know the fate, if not the origins, of the Universe.

Kneedeep in the cosmic overwhelm.
—Diane Ackerman

A torrent of new insights into the nature of the Universe is almost upon us. It is now possible to perform relatively precise calculations on the natures of astronomical objects. The instrumentation for ground-based observations has reached a high state of sophistication. Large space telescopes (LSTs) can be placed above the Earth's atmosphere, where images are steadier, more distant objects can be detected and much finer detail can be resolved. There are new windows in the electromagnetic spectrum: Gamma ray, x-ray, ultraviolet and certain kinds of infrared and radio astronomy are possible for the first time. In the next few decades we will have explored completely the electromagnetic spectrum. It is likely that the enigmas of such mysterious objects as pulsars, quasars, x-ray sources, galactic nuclei and black holes—all of whose most important properties are indetectable in visible light—will be solved.

Astronomy is no longer merely an inferential science, studying light waves emitted or reflected from elsewhere. The advent of space-vehicle exploration permits us to examine the neighboring celestial bodies directly. In the next few decades we should see the nearly complete reconnaissance of the nine planets, 33 moons and miscellaneous comets and asteroids of our solar system. We may see the first solar probe—a spaceship shot into the fiery inferno of the nearest star. And just possibly we may witness the launching of the first space vehicles designed to explore other solar systems.

"We have seen the highest circle of spiraling powers. We have named this circle God. We might have given it any other name we wished: Abyss, Mystery, Absolute Darkness, Absolute Light, Matter, Spirit, Ultimate Hope, Ultimate Despair, Silence."

—Nikos Kazantzakis

CREDITS AND COMMENTS

1: A space vehicle launch from Cape Kennedy. NASA. 2-3: The Milky Way Galaxy edge-on. Mosaic by Lund Observatory, Lund, Sweden. 6-7: The Pleiades, Lick Observatory. 10-11: Apollo photograph, Antarctica to the Mediterranean. NASA. 17: A spiral galaxy in Andromeda NGC891, seen edge-on. Hale Observatories. 19: Copyright © 1973 for VERTEX—The Magazine of Science Fiction by Mankind Publishing Co. Inc.; reprinted by arrangement with the author, Forrest J Ackerman, 2495 Glendower Ave., Hollywood/CA 90027. 20: The Rosetta Stone, in the British Museum, London. Picture collection, N.Y. Public Library. 22: Orlando Busino. 23: Owens Valley Radio Observatory, California Institute of Technology. 24: Extreme ultraviolet image of the Sun, photographed from rocket altitudes. Naval Research Laboratory, Washington. 26-27: The Crab Nebula. Hale Observatories. 28: Albert Einstein. Copyright © 1949 by Philippe Halsman. 30: Joseph Farris. Copyright The New York Times. 32-33: The Museum of Modern Art, New York. van GOGH: "The Starry Night" (1889). 34-35: The Sun in H-alpha radiation. Big Bear Solar Observatory, Big Bear City, Calif. 36-37: Geometry produced by rural agriculture, western France. Institut Geographiqué National, Paris. Photograph from *Worlds From Above,* originally published in German by Reich Publishing Company Ltd., Lucerne/Switzerland. 38-39: Manhattan Island. Environmental series, Administration, U.S. Coast and Geodetic Survey, U.S. Dept. of Commerce. From *Worlds From Above,* op. cit. 40-41: NGC6205 globular star cluster in Hercules. Hale Observatories. 41: The invention of the telescope, Holland, 17th Century. The Bettmann Archive. 42: Drawing by Gahan Wilson. 43: Konstantin Tsiolkovskii, the "father of astronautics," beside an experimental rocket. Sovfoto. 44-45: The Moon and the Earth, to scale, both photographed by Mariner 10. NASA.

46-47: Mount Hadley on the Moon, astronaut and lunar roving vehicle, foreground. Apollo 15 photograph, NASA. 48: The Moon, Apollo 8 photograph. NASA. 50-51: Three views of Mercury during the approach of Mariner 10. NASA. 50: An imagined approach to Mercury in Victorian science fiction. From *Worlds Apart,* pub. by Ferret Fantasy Ltd., 27, Beechcroft Road, Upper Tooting, London SW17; editor: George Locke. 52-53: Mariner 10 photograph of Venus. NASA. 53: An imaginary view of the Venus surface in Victorian science fiction. From *Worlds Apart,* op. cit. 54: An imaginary view of the North Pole of Mars in Victorian science fiction. From *Worlds Apart,* op. cit. 55: A global mosaic of Mariner 9 photographs of Mars. NASA. 56: Deimos from Mariner 9. NASA. 57: Phobos from Mariner 9. NASA. 57: An imaginary view of the moons of Mars in the night sky of the red planet in Victorian science fiction. From *Worlds Apart,* op. cit. 58: An imaginary view of Jupiter and its inner moons from an outer moon in Victorian science fiction. From *Worlds Apart,* op. cit. 59: Jupiter from Pioneer 10. NASA. 60-61: Saturn. New Mexico State University Observatory. Courtesy, Dr. Bradford Smith. 61: *The '?' Motorist* (1906). Picture collection, N.Y. Public Library. 63: Neptune, overexposed. Yerkes Observatory. 64: From "Les Mystères des Infinis" by J. J. Grandville, 1844. 65: Halley's Comet. Hale Observatories. 68-69: Arecibo Observatory at night. National Astronomy and Ionosphere Observatory, Cornell University. 70-71: Mercury from Mariner 10. NASA. 72: Portion of Bosch's Hell in "The Garden of Worldly Delights." Museo del Prado, Madrid. 73: Mariner 10 photograph of detached limb haze above the clouds of Venus. NASA. 74: Ultraviolet photograph of Earth from Skylab. Naval Research Laboratory photograph. Courtesy, George Carruthers.

75: The orbit of the asteroid Toro, between 1800 and 2000, relative to the Sun (center) and the Earth, as shown. Courtesy, P. M. Janiczek, P. K. Seidelmann, and R. L. Duncombe, U.S. Naval Observatory. 76: An illustration for Edgar Rice Burroughs' Mars fiction books. © Ballantine Books. Reprinted by permission of Ballantine Books, a division of Random House, Inc. 77: The Mariner 9 spacecraft. NASA. 78-79:

Courtesy, U.S. Navy. 80: Copyright © 1965 by The New York Times Company. Reprinted by permission. 81: Mariner 9 photograph of Vallis Nirgal on Mars. NASA. 83: Extract from the Ithaca (N.Y.) Daily Journal, showing a segment of Vallis Mangala on Mars, observed by Mariner 9. 84-85: The Martian North Polar Cap. Jet Propulsion Laboratory, Image Processing Laboratory product. From Mariner 9. NASA. 86: The Pavonis Mons volcano on Mars observed by Mariner 9. NASA. 87: Crater-associated wind streaks in Gaea on Mars, observed by Mariner 9. NASA. 88: Percival Lowell at his telescope. Courtesy, Lowell Observatory. 89: Syrtis Major on Mars, observed by C. Huygens (1659) and by A. Dollfus (1956). Courtesy, A. Dollfus, Mendon Observatory, University of Paris. 90: Lemuel Gulliver viewing the floating aerial island of Laputa. Picture collection, N.Y. Public Library. 91: An imaginary floating city in a Martian "canal" in Victorian science fiction. From *Worlds Apart,* op. cit. 93: The U.S. Viking Mars lander. NASA. 94: Orlando Busino. 95: The 200-inch telescope at Mt. Palomar. Ralph Crane for Black Star. 97: Apollo photograph of a great terrestrial storm system. NASA. 98: Jupiter and its Great Red Spot. Lowell Observatory. 100-101: Comet Bennett. Cerro Tololo Inter-American Observatory. 103: Haystack Observatory, Northeastern Radio Observatory Consortium. 104: I. S. Shklovskii. Courtesy, Phyllis Morrison. 106-107: Illustration of the great Moon hoax. The Bettmann Archive. 108: Typical UFO photograph. Library of Congress. 110: James M. and Linda J. Baker, Plattsmouth, Nebraska. Picture made Aug. 10, 1972, at 2:30 p.m. from East side of Jackson Lake facing the Grand Teton Mts. 112-113: *Another Brownie Book* (1890) by Palmer Cox (1840-1924) — he invented the Brownies. 114-115: Easter Island monoliths, readily explicable as the constructs of Easter Island inhabitants. Picture collection, N.Y. Public Library. 116-117: "Long man," 231 feet tall, of Wilmington, England. Note five humans at lower right. Another example of ancient peoples drawing large. © Terence Spencer. 118: From the Tassili frescoes. Picture collection, N.Y. Public Library. 119: Self-portrait. Courtesy, N. J. Z. Sagan. 122: Picture collection, New York Public Library. 124-125: "Horsehead" nebula in Orion. Hale Observatories. 127: Siggs-© Punch-ROTHCO. 128: Pre-Cambrian animal from perhaps a billion years ago, *Tribrachidium heraldicum,* "three-armed shield," about two centimeters in diameter. Courtesy, Prof. M. F. Glaessner, geology dept., University of Adelaide, South Australia, and Centre for Pre-Cambrian Research.
129: Apollo 11 photograph of a human footprint on the Moon. NASA. 131: An experiment in prebiological organic chemistry. Cornell University. 132: Grin and Bear It by Lichty © Field Enterprises Inc. 1973. Courtesy Publishers-Hall Syndicate. 134: Drawing by John Dempsey. 136: From "Un Autre Monde," J. J. Grandville, 1844. 137: From the 1935 motion picture *Things to Come.* Springer/Bettmann Film Archive. 138-139: Confrontation of 18th and 20th Centuries technologies, a sailing vessel and a hovercraft. The New York Times/Neal Boenzi. 143, 145: Copyright © 1974 The New York Times Company. Reprinted by permission. 146: A solar prominence. Courtesy, Sacramento Peak Observatory, Air Force Cambridge Research Laboratory. 147: A radio telescope broadcasting the structure of DNA to space. © 1974, Jon Lomberg. 148: Scanning electron micrograph of Didinium just finishing making dinner of a Paramecium. Courtesy, Gregory Antipa. 149: Nebulosity in NGC 2264 in Monoceros. Hale Observatories. 150: National Radio Astronomy Observatory, Green Bank, W. Va. 152-153: M 16 in Serpens. Lick Observatory. 154: NGC 3031, spiral galaxy in Ursa Major. Hale Observatories. 157: Part of the spiral galaxy M31 in Andromeda. Hale Observatories. 160: Air version of Robinson Crusoe, published in Germany, 1721. The Bettmann Archive.

I am grateful to Steven Soter for a critical reading of the manuscript; to Arthur C. Clarke for introducing me to *Worlds Apart*; and to Diane Ackerman, Cornell University, for permission to reproduce a little of her exquisite astronomical poetry. Part of the text of pp. 12-13 has been published in Harper's Magazine.

The total number of people on Earth today is a little less than four-billion. The total number of human beings who have ever lived is probably seven tens of billions. But the total number of stars in the Milky Way Galaxy is about 250-billion. There are thus ten times as many stars in the Milky Way Galaxy than there are human beings who have ever lived—and there are billions of other galaxies. There are about 10 stars born each year in the Milky Way Galaxy, but there are about 200,000 more people born than die each day on the planet Earth. If the trend continues, the people will soon outnumber the stars.

But the trend cannot continue. Either we will be stupid and destroy ourselves or we will be wise

Drawing by Chas. Addams

"I guess ours isn't the only planet having its troubles."

Drawing by Alan Dunn;

"What is the ultimate truth about ourselves? Various answers suggest themselves. We are a bit of stellar matter gone wrong. We are physical machinery—puppets that strut and talk and laugh and die as the hand of time pulls the strings beneath. But there is one elementary inescapable answer. *We are that which asks the question*. Whatever else there may be in our nature, responsibility towards truth is one of its attributes."

—Sir Arthur Eddington